# Dave Armstrong

# Science and Christianity:
## Close Partners or Mortal Enemies?

© Copyright 2010 by Dave Armstrong

All rights reserved.

Biblical citations are from the Revised Standard Version of the Bible (© 1971) copyrighted by the Division of Christian Education of the National Council of the Churches of Christ in the United States of America. All emphases have been added.

ISBN 978-1-105-53733-2

For related reading on the author's blog, see:

**Philosophy, Science, and Christianity**
http://socrates58.blogspot.com/2006/11/philosophy-christianity-index-page.html

# DEDICATION

To Antoine Lavoisier (1743-1794: the "father of chemistry"), Philippe-Frédéric de Dietrich (1748-1793), Nicolas de Condorcet (1743-1794), Jean Baptiste Gaspard Bochart de Saron (1730-1794), Guillaume-Chrétien de Lamoignon de Malesherbes (1721-1794), and Félix Vicq d'Azyr (1746-1794): scientific martyrs (or probable martyrs, in the cases of Condorcet and d'Azyr) of the so-called "French Enlightenment" and victims of fanatical worshipers of the "goddess of reason" idol.

We have not forgotten you, nor your quite considerable contributions to the worlds of science, philosophy, and mathematics.

**Psalm 19:1** The heavens are telling the glory of God; and the firmament proclaims his handiwork.

**Psalm 111:2** Great are the works of the LORD, studied by all who have pleasure in them.

**Wisdom of Solomon 11:20** . . . But thou hast arranged all things by measure and number and weight.

**Proverbs 25:2** It is the glory of God to conceal things, but the glory of kings is to search things out.

**Acts 17:28** . . . In him we live and move and have our being . . .

**Romans 1:19-20** For what can be known about God is plain to them, because God has shown it to them. Ever since the creation of the world his invisible nature, namely, his eternal power and deity, has been clearly perceived in the things that have been made. So they are without excuse;

**Colossians 1:16-17** . . . all things were created through him and for him. He is before all things, and in him all things hold together.

**Hebrews 1:3** . . . upholding the universe by his word of power ...

# CONTENTS

Dedication..........................................................................3

1. Christianity's Central Role in the Conception and Development of Modern Science.........................................7

2. Fertile Soil and Roots of Modern Science: 33 Prominent Christians Prior to 1000 A. D. With Empiricist, Proto-Scientific Views...................................................................................19

3. 59 Catholic Medieval and/or Scholastic Proto-Scientists From 1000 to 1500 A. D. ...................................................37

4. 70 Catholic, Protestant and Otherwise Religious Prominent Scientists: 1500-1700 (From Copernicus to Steno, Boyle, and Ray)...................................................................................73

5. 36 Catholic, Protestant and Otherwise Religious Prominent Scientists: 1700-1800 (From Newton to Linnaeus, Boscovich, and Lavoisier).....................................................................107

6. 41 Catholic, Protestant and Otherwise Religious Prominent Scientists: 1800-1850 (From Dalton to Humboldt, Cuvier, and Faraday)................................................................................127

7. 56 Catholic, Protestant and Otherwise Religious Prominent Scientists: 1850-1900 (From Maxwell to Mendel, Pasteur, and Kelvin)..................................................................................153

8. 31 Catholic, Protestant and Otherwise Religious Prominent Scientists: 1900-1950 (From Einstein to Planck, Eddington, and Lemaître)..................................................................191

9. 115 Scientific Fields of Study Founded or Extraordinarily Advanced by Christian or Theistic Scientists / 34 Prominent Catholic Priest-Scientists and Mathematicians: 1500-1950..................................................................................221

10. Albert Einstein's "Cosmic Religion"..........................231

11. The Galileo Case: Historical Facts and Neglected Considerations vs. Secular Revisionist Myths....................249

12. Galileo and Other Prominent 16th-17th Century Astronomers' Acceptance of Astrology...........................257

13. "No One's Perfect": Scientific Errors of Galileo and 16th-17th Century Cosmologies Rescued From Inexplicable Obscurity............................................................................269

14. The Execution of Antoine Lavoisier: the Great Catholic Scientist and "Father of Chemistry" by "Enlightened" French Revolutionaries.................................................................275

15. Christian Influence on Science: Master List of Scores of Bibliographical and Internet Resources..........................285

# Chapter One

# Christianity's Central Role in the Conception and Development of Modern Science

[**Preliminary note on sources**: I shall, throughout the biographical chapters 2-8 (326 "mini-biographies" in all), massively cite articles from the online encyclopedia, Wikipedia: mostly biographies of scientists; and also from *The Catholic Encyclopedia*: published in 1913 and available online in its entirety, due to the guidance and tireless work of Kevin Knight. The Wikipedia articles themselves, in biographical entries, quite often draw heavily from *The Catholic Encyclopedia*. I cite both sources so persistently that I don't bother to make quotation marks. But readers should be aware that in those seven chapters I am drawing mostly from these sources and some additional ones listed in the bibliography in chapter 15, with a few changed or added words on my part, and editing for length and focusing purposes. In each case I make a link to the article I am using. Wikipedia content is allowed very wide usage, according to the Creative Commons Attribution-ShareAlike 3.0 Unported License and the GNU Free Documentation License: as explained in great detail on its Wikipedia:Copyrights page. *The Catholic Encyclopedia* is in the public domain, due to its age.

Secondly, by its very nature, this work depends in great measure on the marvelous Internet links technology, in order to make instantly accessible, articles on any given scientific (and also mathematical and sometimes philosophical) matter or discovery. If readers want to know more about the particulars referred to in passing, then the many hundreds of links provided in this work will be quite convenient for that purpose. My own greatest interest lies in the areas of history of science and philosophy of science: and it is to those special areas that this book is devoted in the larger sense.]

It's very fashionable nowadays for atheists (including atheist scientists) to make extreme claims about the alleged utter incompatibility between Christianity and science. It is said that the two are antithetical, or that God was ruled out of science or disproven by scientific findings (particularly Darwinian evolution) long ago, or that science proceeds forward based on reason and evidence, whereas religion (being faith-based) supposedly has no reason and cares little or nothing for evidence, or that one cannot consistently be a Christian and also a "real" scientist.

Mano Singham, an adjunct associate professor of physics at Case Western Reserve University and author of *God vs. Darwin: The War Between Evolution and Creationism in the Classroom* (Rowman & Littlefield, 2009), quintessentially exhibited this mentality in an online article:

> [T]he fact that some scientists are religious is not evidence of the compatibility of science and religion. As Michael Shermer, founder and editor of *Skeptic* magazine, says in his book *Why People Believe Weird Things* (A. W. H. Freeman / Owl Book, 2002), "Smart people believe weird things because they are skilled at defending beliefs they arrived at for non-smart reasons." Jerry Coyne, a professor in the department of ecology and evolution at the University of Chicago, notes, "True, there are religious scientists and Darwinian churchgoers. But this does not mean that faith and science are compatible, except in the trivial sense that both attitudes can be simultaneously embraced by a single human mind."

("The New War Between Science and Religion," *The Chronicle of Higher Education*, 9 May 2010)

Statements along these general lines could be multiplied *ad infinitum, ad nauseum*. It's been this way, unfortunately, for quite some time. The famous and influential philosopher Bertrand Russell (1872-1970) proclaimed in his essay *Why I am Not a Christian* (1927):

> Science can teach us, and I think our own hearts can teach us, no longer to look around for imaginary supports, no longer to invent allies in the sky, but rather to look to our own efforts here below to make this world a better place to live in, instead of the sort of place the churches in all these centuries have made it.

Social critic and rapier wit H. L. Mencken (1880-1956) informed us of how anti-scientific Christians supposedly are, in several ridiculous utterances:

> A man full of faith is simply one who has lost (or never had) the capacity for clear and realistic thought. He is not a mere ass: he is actually ill.

> Religion is fundamentally opposed to everything I hold in veneration -- courage, clear thinking, honesty, fairness, and, above all, love of the truth.

> The Christian church, in its attitude toward science, shows the mind of a more or less enlightened man of the Thirteenth Century. It no longer believes that the earth is flat, but it is still convinced that prayer can cure after medicine fails.

> The believing mind is externally impervious to evidence. The most that can be accomplished with it is to induce it to substitute one delusion for another. It rejects all overt evidence as wicked.

Various permutations of these ultra-intolerant, downright prejudiced, condescending themes are observed all the time in agnostic / atheist, scientific, and pseudo-scientific circles. Documenting even a thousandth of them would fill up a volume thicker than the most in-depth dictionary.

I shall contend in what follows, that not only are science and Christianity *compatible*, but that modern science would not have even *gotten off the ground* if it hadn't been for medieval, scholastic, Catholic thought for the previous several hundred years: in the realm of empiricism and scientific observation.

Nor would science have been consistently *sustained* over the last 500 years in its fabulous discoveries and breathtaking progress, since this was also overwhelmingly fueled by Christian and otherwise theistic scientists.

The foundations of modern science (once it *did* get off the ground in the 16th century) were overwhelmingly Christian. Other important scientists through the years had religious views that were sub-theistic, such as deism or pantheism (for example, Einstein). To say that science and religion are fundamentally incompatible is literally a nonsensical statement that would obliterate science at its very roots and presuppositions and bedrock premises. It's a self-defeating proposition, and is "historically illiterate" to even propose such a ludicrous notion.

One cannot exist only in the present. Present-day science didn't jump out of a vacuum, fully developed. It has a history of assumptions that were built upon and expanded. These were unable to be separated from Christianity. Since they are intrinsic to the scientific enterprise, it is is impossible now to attempt to separate science and Christianity altogether, as if the past and the history of science was not what it was. Oftentimes, scientists (and atheists who wax eloquently and dogmatically about science) are as historically uninformed as they are unacquainted with philosophy in general or the philosophical roots of science itself.

Modern science, in order to function and proceed at all, had to accept several unproven axioms. And these axioms were essentially derived from Christianity. Eminent physicist Paul Davies (as far as I can tell, an Einstein-like pantheist, but not a theist) makes the basic, introductory-type observations in this regard:

[S]cience has its own faith-based belief system. All science proceeds on the assumption that nature is ordered in a rational and intelligible way. You couldn't be a scientist if you thought the universe was a meaningless jumble of odds and ends haphazardly juxtaposed....

... to be a scientist, you had to have faith that the universe is governed by dependable, immutable, absolute, universal, mathematical laws of an unspecified origin. You've got to believe that these laws won't fail, that we won't wake up tomorrow to find heat flowing from cold to hot, or the speed of light changing by the hour....

Clearly, then, both religion and science are founded on faith — namely, on belief in the existence of something outside the universe, like an unexplained God or an unexplained set of physical laws, maybe even a huge ensemble of unseen universes, too. For that reason, both monotheistic religion and orthodox science fail to provide a complete account of physical existence.

This shared failing is no surprise, because the very notion of physical law is a theological one in the first place, a fact that makes many scientists squirm. Isaac Newton first got the idea of absolute, universal, perfect, immutable laws from the Christian doctrine that God created the world and ordered it in a rational way. Christians envisage God as upholding the natural order from beyond the universe, while physicists think of their laws as inhabiting an abstract transcendent realm of perfect mathematical relationships.

And just as Christians claim that the world depends utterly on God for its existence, while the converse is not the case, so physicists declare a similar asymmetry: the universe is governed by eternal laws (or meta-laws), but the laws are completely impervious to what happens in the universe....

In other words, the laws should have an explanation from within the universe and not involve appealing to an external agency. The specifics of that explanation are a matter for future research. But until science comes up with a testable theory of the laws of the universe, its claim to be free of faith is manifestly bogus.

("Taking Science on Faith," *New York Times*, 24 November 2007)

He expressed similar thoughts in his 1995 Templeton Prize Address:

It was from the intellectual ferment brought about by the merging of Greek philosophy and Judaeo-Islamic-Christian thought, that modern science emerged, with its unidirectional linear time, its insistence on nature's rationality, and its emphasis on mathematical principles. All the early scientists such as Newton were religious in one way or another. They saw their science as a means of uncovering traces of God's handiwork in the universe. What we now call the laws of physics they regarded as God's abstract creation: thoughts, so to speak, in the mind of God. So in doing science, they supposed, one might be able to glimpse the mind of God. . . . science can proceed only if the scientist adopts an essentially theological world view. . . .

Philosopher Alfred North Whitehead (1861-1947) also saw this clearly in 1925:

In the first place, there can be no living science unless there is a widespread instinctive conviction in the existence of an *Order Of Things*. And, in particular, of an *Order Of Nature* . . . The inexpugnable belief that every detailed occurrence can be correlated with its antecedents in a perfectly definite manner . . . must come from the medieval insistence on the rationality of God . . .

My explanation is that the faith in the possibility of science, generated antecedently to the development of modern scientific theory, is an unconscious derivative from medieval theology.

The faith in the order of nature that has made possible the growth of science is a particular example of a deeper faith. This faith cannot be justified by any inductive generalisation. It springs from direct inspection of the nature of things as disclosed in our immediate present experience.

(*Science and the Modern World*; reprinted by Free Press, 1997, 3-4, 13, 18)

One of the leading philosophers of science, Thomas Kuhn (1922-1996), elucidated the medieval background in his book, *The Copernican Revolution* (New York: Vintage Books / Random House, 1959):

After the Dark Ages the Church began to support a learned tradition as abstract, subtle, and rigorous as any the world has known . . . The Copernican theory evolved within a learned tradition sponsored and supported by the Church . . . (p. 106)

The centuries of scholasticism are the centuries in which the tradition of ancient science and philosophy was simultaneously reconstituted, assimilated, and tested for adequacy. As weak spots were discovered, they immediately became the foci for the first effective research in the modern world. The great new scientific theories of the sixteenth and seventeenth centuries all originate from rents torn by scholastic criticism in the fabric of Aristotelian thought. Most of those theories also embody key concepts created by scholastic science. And more important than these is the attitude that modern scientists inherited from their medieval predecessors: an unbounded faith in the power of human reason to solve the problems of nature. (p. 123)

Historian of science James Hannam, in his marvelously informative treatise, "Christianity and the Rise of Science," stated:

> I have often come across anti Christians who simply cannot bring themselves to accept that Christianity had anything to do with the development of their beloved science. There are, I think, two reasons for this. First, they have fed themselves an unrelenting diet of nineteenth century anti religious myths like those found in Andrew Dickson White's *The Warfare of Science and Theology* and John William Draper's *History of the Conflict between Religion and Science* so they cannot bear to admit a single good thing has come from Christianity despite all the evidence around them.
>
> Others have felt that any discussion on science and religion is killed stone dead by simply mentioning the unfortunate but, in the long term, not very significant Galileo affair. . . . The second problem is that the history of science as an academic subject is still in its infancy and medieval science, which I believe is the vital period, is even more neglected due to the lack of Latin language skills.. . .
>
> The early modern scientists were inspired by their faith to make their discoveries and saw studying the creation of God as a form of worship. . . . For the anti Christians desperate not to give credit for their own faith of scientism to the religion they hate, two questions must be answered. First, if the dominant world view of medieval Europe was as hostile to reason as they would like to suppose, why was it here rather than anywhere else that science arose? And secondly, given that nearly every one of the founders and pre founders of science were unusually devout (although not always entirely orthodox) even by the standards of their own time, why did they make the scientific breakthroughs rather than their less religiously minded contemporaries? I wonder if I will receive any answers.

Dr. Francis S. Collins (a former atheist) is Director of the National Human Genome Research Institute at the National Institute of Health. He leads the Human Genome Project (mapping and sequencing human DNA, and specifically determining function). He has identified the genes responsible for cystic fibrosis, neurofibromatosis, Huntington's disease and Hutchison-Gilford progeria syndrome. In an interview with Bob Abernethy of PBS, he stated:

> I think there's a common assumption that you cannot both be a rigorous, show-me-the-data scientist and a person who believes in a personal God. I would like to say that from my perspective that assumption is incorrect; that, in fact, these two areas are entirely compatible and not only can exist within the same person, but can exist in a very synthetic way, and not in a compartmentalized way. I have no reason to see a discordance between what I know as a scientist who spends all day studying the genome of humans and what I believe as somebody who pays a lot of attention to what the Bible has taught me about God and about Jesus Christ. Those are entirely compatible views. . . .
>
> They coexist. They illuminate each other. And it is a great joy to be in a position of being able to bring both of those points of view to bear in any given day of the week. The notion that you have to sort of choose one or the other is a terrible myth that has been put forward, and which many people have bought into without really having a chance to examine the evidence.

Loren Eiseley (1907-1977), an anthropologist, educator, philosopher, and natural science writer, who received more than 36 honorary degrees, and was himself an agnostic in religious matters, observed:

> It is the Christian world which finally gave birth in a clear articulated fashion to the experimental method of science itself . . . It began its discoveries and made use of its

method in the faith, not the knowledge, that it was dealing with a rational universe controlled by a Creator who did not act upon whim nor inference with the forces He had set in operation. The experimental method succeeded beyond man's wildest dreams but the faith that brought it into being owes something to the Christian conception of the nature of God. It is surely one of the curious paradoxes of history that science, which professionally has little to do with faith, owes its origins to an act of faith that the universe can be rationally interpreted, and that science today is sustained by that assumption.

(*Darwin's Centenary: Evolution and the Men who Discovered it*, New York: Doubleday: 1961, 62)

The prominent British philosopher Robin George Collingwood (1889-1943) wrote in similar fashion:

The presuppositions that go to make up this Catholic faith, preserved for many centuries by the religious institutions of Christendom, have as a matter of historical fact been the main or fundamental presuppositions of natural science ever since.

(*Essay on Metaphysics*, Oxford University Press: 1940, 227)

H. Floris Cohen, in his book, *The Scientific Revolution: A Historiographical Inquiry* (University of Chicago Press, 1994), described the views of science historian Benjamin Farrington (1891-1974), in his works, *Science in Antiquity* (1936) and *Greek Science* (1949):

The other novel element that contributed crucially to the changed atmosphere in which the heritage of Greek science was received in western Europe is the biblical world-view, This entailed a more positive appreciation of labor, of the arts, and of the possibility of the amelioration of man's future fate generally. . . . Greek science . . . was revitalized both by the achievements of medieval technology and by an optimistic, active world-view derived from the Bible. (pp. 248-249)

Eminent German physicist and philosopher Carl Friedrich von Weizsäcker (1912-2007), in his book, *The Relevance of Science* (New York: Harper and Row: 1968, 163), even went so far as to conclude that modern science is a "legacy, I might even have said, a child of Christianity."

Lastly, among many other comments that could be produced along these lines, the Nobel Prize-winning biochemist Melvin Calvin (1911-1997), referring to the idea that the universe has a rational order, wrote:

> As I try to discern the origin of that conviction, I seem to find it in a basic notion . . . enunciated first in the Western world by the ancient Hebrews: namely, that the universe is governed by a single God, and is not the product of the whims of many gods, each governing his own province according to his own laws. This monotheistic view seems to be the historical foundation for modern science.
>
> (*Chemical Evolution*, Oxford: Clarendon Press, 1969, 258)

## Chapter Two

## Fertile Soil and Roots of Modern Science: 33 Prominent Christians Prior to 1000 A. D. With Empiricist, Proto-Scientific Views

> In a momentous move, Clement of Alexandria (ca. 150-ca. 215) and his disciple Origen of Alexandria (ca. 185-ca. 254) laid down the basic approach that others would follow. Greek philosophy was neither inherently good nor bad, but was one or the other depending on how it was used by Christians. . . . Philosophy and science could be studied as "handmaidens to theology" . . . The handmaiden concept of Greek learning was widely adopted and became the standard Christian attitude toward secular learning. That Christians chose to accept pagan learning within limits was a momentous decision. . . . Their education was heavily infiltrated by Latin and Greek pagan literature and philosophy. . . . Christians realized that they could not turn their backs on Greek learning.
>
> (Edward Grant, *The Foundations of Modern Science in the Middle Ages: Their Religious, Institutional and Intellectual Contexts* [Cambridge, 1996], pp. 3-4)

On a spectrum of *pagan* values, from cosmic religion to Gnostic repudiation of the cosmos, the church fathers chose a middle position. There can be no doubt that biblical teaching about the creation as God's handiwork was influential in determining where on the spectrum Christians would land, and therefore it is clear that their Christianity was highly relevant to the issue . . . It seems unlikely, therefore, that the advent of Christianity did anything to diminish the support given to scientific activity or the number of people involved in it.

(David Lindberg, in Lindberg and Ronald Numbers, editors, *God and Nature: Historical Essays on the Encounter Between Christianity and Science* [Univ. of California Press, 1986], pp. 32-33)

Other historians have pointed out that although the Church Fathers subordinated the claims of science to the demands of faith, they rarely rejected Greco-Roman science and philosophy outright. . . . The dominant position was somewhere between that of Tertullian and Gregory [Nazianzus]: science and human reason in general were useful to the Christian as long as knowledge about the natural world never supplanted love of God or contradicted the tenets of religious belief.

(Elspeth Whitney, *Medieval Science and Technology* [Greenwood, 2004], p. 5)

[T]he vast majority of Christian leaders looked favorably on the Greco-Roman medical tradition, viewing it as a divine gift, an aspect of divine providence, the use of which was legitimate and perhaps even obligatory. Basil of Caesarea (ca. 330-79) spoke for many of the church fathers when he wrote that "we must take great care to employ this medical art, if it should be necessary . . ."
    [H]ow did the presence and influence of the Christian church affect knowledge of, and attitudes toward, nature? The standard answer, developed in the eighteenth and nineteenth centuries and widely propagated in the twentieth, maintains that Christianity presented serious obstacles to the advancement of science

and, indeed, sent the scientific enterprise into a tailspin from which it did not recover for more than a thousand years. The truth, as we shall see, is dramatically different, far more complicated, and a great deal more interesting...

Naturally enough, the kind and level of education and intellectual effort favored by the church fathers was that which supported the mission of the church as they perceived it. But this mission, interestingly, did not include the suppression of scientific investigations and ideas.

If we compare the early church with a modern research university or the National Science Foundation, the church will prove to have failed abysmally as a supporter of science and natural philosophy. But such a comparison is obviously unfair. If, instead, we compare the support given to the study of nature by the early church with the support available from any other contemporary social institution, it will become apparent that the church was the major patron of scientific learning. Its patronage may have been limited and selective, but limited and selective patronage is a far cry from opposition.

The contribution of the religious culture of the early Middle Ages to the scientific movement was thus primarily one of preservation and transmission. The monasteries served as the transmitters of literacy and a thin version of the classical tradition (including science or natural philosophy) through a period when literacy and scholarship were severely threatened. Without them, Western Europe would not have had more science, but less.

(David Lindberg, *The Beginnings of Western Science* [Univ. of Chicago Press, 2nd ed., 2008], pp. 325 and 148-150, 156-157)

**St. Clement of Rome** (d. c. 101; pope) He accepted a good deal of Greek mathematics and astronomy, including belief that the earth was spherical. Unlike Aristotle, however, for him the earth was not eternal and it was sharply distinguished from the divine. Both the heavens and the earth were created and they were orderly: "the sun, the moon and the dancing stars . . . circle in harmony within the bounds assigned to them." [source: Tripp]

**St. Clement of Alexandria** (c. 150-c. 215)

> [T]he catechetical school of Alexandria . . . acquired the character of a Christian academy in which all Greek science was studied and made to do apologetic service in favour of the Christian cause. Under Clement and Origen it reached the acme of its renown . . .
>
> (Otto Bardenhewer and Thomas Joseph Shahan, *Patrology: The Lives and Works of the Fathers of the Church* [Herder, 1908], p. 127; see also the citation from Edward Grant at the top of this chapter)

**Origen** (c. 185-c. 254) He accepted the sphericity of the earth [source: Hannam: "The Myth of the Flat Earth"] See also the previous entry and citation from Edward Grant at the top of this chapter.

**St. Gregory Nazianzus** (329-389; bishop and Doctor of the Church) He wrote:

> For as we ought not to neglect the heavens, and earth, and air, and all such things, because some have wrongly seized upon them, and honour God's works instead of God: but to real what advantage we can from them for our life and enjoyment . . . from the works of nature apprehending the Worker . . . as we know that neither fire, nor food, nor iron, nor any other of the elements, is of itself most useful, or most harmful, except according to the will of those who use it; and as we have compounded

healthful drugs from certain of the reptiles; so from secular literature we have received principles of enquiry and speculation . . .

(David Lindberg, in Lindberg and Ronald Numbers, editors, *God and Nature: Historical Essays on the Encounter between Christianity and Science* [Univ. of California Press, 1986], p. 29; citing *The Panegyric on St. Basil*: from the Schaff / Wace 38-volume translation of the Church fathers, vol. 7, 398-399)

<u>St. Basil the Great</u> (c. 330-379; bishop and Doctor of the Church) In contrast to Aristotle, he believed the heavens and the earth were made up of the same materials: earth, air, fire and water, and also questioned the Aristotelian view that divine spirits in the heavenly bodies must continue imparting motion directly to everything that moves. By analogy with a child's top, he spoke of the heavenly bodies, "which after the first impulse, continue their revolutions, turning upon themselves when once fixed in their centre; thus nature, receiving the impulse of this first command, follows without interruption the course of the ages". Basil's spinning top provides an early formulation of the idea of impetus. His views on creation allow for the principle of the conservation of momentum, or of inertia, that appeared repeatedly in Christian thinkers over the next twelve centuries. [source: Tripp] He believed in a spherical earth. [source]

"If you observe carefully the members even of the animals, you will find that the Creator has added nothing superfluous, and that He has not omitted anything necessary." He drew lessons from the migration of fish, the stealth of the octopus, the function of the elephant's trunk, the behavior of dogs tracking wild animals, and the existence of both poisonous and edible plants. All play their designated role in nature, even poisonous plants, for as Basil argued, "there is no one plant without worth, not one without use. Either it provides food for some animal, or has been sought out for us by the medical profession for the relief of certain diseases." Thus did Basil respond to those who wondered why God would create poisonous plants capable of killing humans.

(Edward Grant, *The Foundations of Modern Science in the Middle Ages: Their Religious, Institutional and Intellectual Contexts* [Cambridge, 1996], p. 6; primary sources unable to be accessed in Google Books)

**St. Gregory of Nyssa** (c. 335-c. 394; bishop) He wrote about the spherical earth:

> As, when the sun shines *above* the earth, the shadow is spread over its lower part, because its spherical shape makes it impossible for it to be clasped all round at one and the same time by the rays, and necessarily, on whatever side the sun's rays may fall on some particular point of the globe, if we follow a straight diameter, we shall find shadow upon the opposite point, and so, continuously, at the opposite end of the direct line of the rays shadow moves round that globe, keeping pace with the sun, so that equally in their turn both the upper half and the under half of the earth are in light and darkness.
>
> (*On the Soul and the Resurrection*)

**St. Ambrose** (c. 336-397; bishop and Doctor of the Church) He believed in a spherical earth. [source: Hannam: "The Myth of the Flat Earth"] [source]

> [H]is work is full of observation of the facts of the natural world. He has minute descriptions of quails and storks and swallows, of bees and crickets, of trees and their modes of reproduction, of evaporation and the action of rain, of human anatomy and physiology.
>
> (Edward K. Rand, *Founders of the Middle Ages* [New York: Dover Puiblications, 1957; originally 1928, 93)

**St. Jerome** (c. 343-420; priest and Doctor of the Church) He referred to the earth as a "sphere" in his Letter 124 to Avitas (5).

**St. Augustine** (354-430; bishop and Doctor of the Church) He was the dominant thinker of the first thousand years of Christian history. For him, the universe, being the creation of God, was not eternal but finite in space and time. Time itself had its created beginning. . . . The Greek notion of cyclic returns was ridiculous, and eliminated the possibility of happiness. [source: Tripp] He wrote, about science and Scripture:

> Usually, even a non-Christian knows something about the earth, the heavens, and the other elements of this world, about the motion and orbit of the stars and even their size and relative positions, about the predictable eclipses of the sun and moon, the cycles of the years and the seasons, about the kinds of animals, shrubs, stones, and so forth, and this knowledge he hold to as being certain from reason and experience. Now, it is a disgraceful and dangerous thing for an infidel to hear a Christian, presumably giving the meaning of Holy Scripture, talking nonsense on these topics; and we should take all means to prevent such an embarrassing situation, in which people show up vast ignorance in a Christian and laugh it to scorn. The shame is not so much that an ignorant individual is derided, but that people outside the household of faith think our sacred writers held such opinions, and, to the great loss of those for whose salvation we toil, the writers of our Scripture are criticized and rejected as unlearned men. If they find a Christian mistaken in a field which they themselves know well and hear him maintaining his foolish opinions about our books, how are they going to believe those books in matters concerning the resurrection of the dead, the hope of eternal life, and the kingdom of heaven, when they think their pages are full of falsehoods and on facts which they themselves have learnt from experience and the light of reason? Reckless and incompetent expounders of Holy

Scripture bring untold trouble and sorrow on their wiser brethren when they are caught in one of their mischievous false opinions and are taken to task by those who are not bound by the authority of our sacred books. For then, to defend their utterly foolish and obviously untrue statements, they will try to call upon Holy Scripture for proof and even recite from memory many passages which they think support their position, although they understand neither what they say nor the things about which they make assertion.

(From: *The Literal Meaning of Genesis* [*De Genesi ad litteram libri duodecim*], Bk. I, ch. 19, 39: translation by J. H. Taylor, in *Ancient Christian Writers* [Newman Press, 1982], volume 41, pp. 42-43)

In the same work (p. 33), Augustine refers to the spherical earth:

[T]here was nothing to prevent the massive watery sphere from having day on one side by the presence of light, and on the other side, night by the absence of light. Thus, in the evening, darkness would pass to that side from which light would be turning to the other. (Bk. I, ch. 12, 25)

And he wrote about the moon being illuminated by the sun:

[I]t is not the heavenly body itself that changes but the illumined surface of it. . . . it is always full, though it does not always appear so to the inhabitants of earth. The same explanation holds even if it is illumined by the rays of the sun. . . . It is only when it is opposite the sun that the whole of its illuminated surface is visible to us. (Bk. II, ch. 11, 31, pp. 68-69)

He rejected astrology (something that Galileo and Kepler still had not done 1200 years later). partially on the basis of observation of twins:

> [L]et us . . . wholeheartedly reject all subtleties of astrologers and their so-called scientific observations . . . which they fancy established by their theories. . . . with headstrong impiety they treat evil-doing that is justly reprehensible as if God were to blame as the maker of the stars, and not man as the author of his own sins. . . .
>
> What more absurd and stupid than, after assenting to the foregoing argument, to say that the influence and power of stars is only over the lives of men? Such a theory is refuted by the case of those twins that spend their lives in different circumstances, one prosperous and the other wretched . . .(Bk. II, ch. 17, 35-36, p. 71)

John F. McCarthy elaborates on St. Augustine's positing of something not unlike theistic evolution:

> This theory of primordial packages of forms later to emerge (often referred to by commentators as "seminal reasons") is certainly developmental, but does not correspond with Darwinian evolution. Essential to Augustine's theory is the idea that the order later to emerge was instilled by God in the beginning. Augustine also requires subsequent interventions by God to "plant" the forms whose "numbers" had already been instilled. Thus, as St. Thomas [Aquinas] points out, the ability of the earth to produce living forms was visualized by Augustine as a passive potency which disposed the matter to receive the forms but did not create the forms themselves. Augustine's theory of primordial packages deserves more ample meditation and analysis in another place, especially with reference to theories of the development of living things, . . . Genesis 1:6-8 witnesses in several ways to the creative action of God. As the divine Fashioner of the universe, God guided the energies

that He had invested in the primal matter by his creative intervention on the first day to bring the cosmos to its structured state. This is the unfolding of the active potency contained in St. Augustine's "primordial packages." But there is also implied in these verses an upward progress in the order of inorganic being which seems to have required additional creative divine interventions.

(A Neo-Patristic Return to the First Four Days of Creation, Part IV; see also Parts One, Two, Three, Five, and Six)

See also, "How Augustine Reined in Science," Kenneth J. Howell (*This Rock*, March 1998); Davis A. Young, "The Contemporary Relevance of Augustine's View of Creation," and Andrew J. Brown, "The Relevance of Augustine's View of Creation Re-Evaluated" (PDF).

**Paulus Orosius** (c. 375-c. 418; priest) He believed in a spherical earth. [source]

**Pope St. Leo the Great** (c. 400-461; Doctor of the Church) He wrote:

> Use these visible creatures as they ought to be used, as you use earth, sea, sky, air, springs and rivers; and praise and glorify the Creator for everything fair and wonderful in them . . . We are not, of course, . . . telling you this to persuade you to despise the works of God, or to think that there is anything against your faith in the things which the good God has made good; but so that you may use every kind of creature, and all the furniture of this world, reasonably and temperately.

(David Lindberg, in Lindberg and Ronald Numbers, editors, *God and Nature: Historical Essays on the Encounter between Christianity and Science* [Univ. of

California Press, 1986], 32; citing *In nativitate Domini sermo VII*)

**Theodosius II** (401-450) A non-literary but graphic indication that people in the Middle Ages believed that the Earth was a sphere, is the use of the *orb* (globus cruciger) in the regalia of many kingdoms and of the Holy Roman Empire. It is attested from the time of the Christian late-Roman emperor Theodosius II (423) throughout the Middle Ages. [source: Wikipedia: "Flat Earth"]

**Procopius of Gaza** (c. 465-528) He argued against life in the antipodes of the earth (the other side of the earth). In so doing, he presupposed that the earth was a sphere. [source: White, vol. 1, 104]

**Anthemius of Tralles** (c. 474-before 558) He collaborated with Isidore of Miletus to build the church of Hagia Sophia in Constantinople (what is today Istanbul in Turkey): the largest cathedral in the world for nearly a thousand years. He was also a capable mathematician. He described the string construction of the ellipse and he wrote a book on conic sections. Anthemius assumes a property of an ellipse not found in Apollonius work, that the equality of the angles subtended at a focus by two tangents drawn from a point, and having given the focus and a double ordinate he goes on to use the focus and directrix to obtain any number of points on a parabola—the first instance on record of the practical use of the directrix. [source: Wikipedia bio]

**Isidore of Miletus** (6th c.) He was one of the two architects (the other being Anthemius of Tralles) who designed the church of Hagia Sophia in Constantinople. He was also an able mathematician, to him we owe the T-square and string construction of a parabola. [source: Wikipedia bio]

**Boethius** (c. 480-524) He documented several ideas relating science to music, including suggestions of sound waves (comparing sound to the waves made by throwing a stone into the water) and the notion that pitch is related to the physical property of frequency:

> [W]hen air is struck and produces a sound, it impels other air next to it and in a certain way sets a rounded wave of air in motion, and is thus dispersed and strikes simultaneously the hearing of all who are standing around.
>
> [T]he same string, if it is tightened further, gives a higher pitched sound; if it is loosened, a sound of lower pitch. For when it is tighter it renders a swifter impulse and returns more quickly, striking the air more frequently and more densely.

[sources: *Encyclopedia Britannica*; Marshall Clagett, *Greek Science in Antiquity* (Dover Pub., 2002, pp. 74, 151)]

He regarded the Earth as a sphere (*globus terrae*) in the center of a spherical cosmos, in his influential, and widely translated, *Consolation of Philosophy*. [source: Wikipedia: "Flat Earth"]

**Cassiodorus** (c. 485-c. 585; monk) He believed in a spherical earth. [source]

**John Philoponus** (c. 490-c. 570; aka John of Alexandria) [Monophysite] He was a professor in the school of philosophy in Alexandria, and the first to mount a devastating critique of the deductive method and much of the content of Aristotle's physics and cosmology. There was no rival to its thoroughness until Galileo. For him, heavenly bodies were not animated beings, but were made of the same stuff as this world. The light from the stars was the same as that of glow-worms and luminescent fish. Astrology was rejected as pagan. Similarly, the heavenly bodies were not perfect. They did not move with regularity in the perfect shape of the circle - a simple matter of observation. The apparent changelessness of the universe did not mean that it is eternal. It had a beginning and will have an end. He rejected Aristotle's view that heavier bodies fall faster than lighter ones (a thousand years before Galileo!). He declared, "Our view may be corroborated by actual observation more effectively than by any sort of verbal argument." His theories on motion were the forerunner of the later theories of inertia and momentum that are embedded in Newton's first law of motion. As regards nature, he stated that God, having finished the creation of the universe, "hands over to nature the generation of the elements one out of another, and the generation of the rest out of the elements." That sounds like a summary of the evolution of the universe from basic materials that modern science would identify with. The relative autonomy of nature, with its own order and laws, is basic to science, and these early Christian thinkers were laying the foundations. . . . Galileo knew the key work of Philoponus, from a thousand years earlier. [source: Tripp] He accepted the sphericity of the earth [source: Hannam: "The Myth of the Flat Earth"] He understood light as something dynamical and that light and heat may best be explained as consequences of the nature of the sun, which is fire (*In Meteor.* 49). Heat is generated when the rays emanating from the sun are refracted and warm the air through friction. He concluded, against Aristotle, that there is in fact nothing to prevent one from imagining motion taking place through a void. As regards the natural motion of bodies falling through a medium, Aristotle's verdict that the speed is proportional to the weight of the moving bodies and indirectly proportional to the density of the medium is disproved by Philoponus through appeal to the same kind of experiment that Galileo was to carry out centuries later (*In Phys.* 682-84). Philoponus insisted that a clear conception of the void is not only coherent but also necessary if one wants to explain movement in a plenum. When bodies move and in consequence exchange places, this presupposes that somehow there is empty space available to be filled by them (*In Phys.* 693f.). Again, there are certain phenomena which clearly exhibit the force of the vacuum, for example handling a pipette (*clepsudra*), which allows one to raise small quantities of fluids, or the fact that one can suck up water through a pipe (*In Phys.* 571f). Philoponus' elaborate defense of the void (*In Phys.* 675-94) is closely related to his conceptions of place and space (*In Phys.* 557-85). The *De opificio mundi* has received some attention from historians of science, because Philoponus suggests at one point (I 12) that the movement of the heavens could be explained by a 'motive force' impressed on the celestial bodies by God at the time of creation. Philoponus compares the rotation implanted in the celestial bodies to the rectilinear movements of the elements as well as to the movements of animals: curiously, these are all understood as natural motions that are due to the creator's divine impetus. In virtue of this bold suggestion Philoponus is often credited with having envisaged, for the first time, a unified theory of dynamics, since he strove to give the same kind of explanation for phenomena which Aristotle had to explain by different principles, depending upon their different cosmological contexts. [source: *Stanford Encyclopedia of Philosophy*: "John Philoponus"]

**St. Isidore of Seville** (c. 560-636; archbishop and Doctor of the Church) He believed in a spherical earth. [source] [second source: Hannam: "The Myth of the Flat Earth"] He taught in his widely read encyclopedia, the *Etymologies*, that the Earth was round and "resembles a wheel". This was widely interpreted as referring to a flat disc-shaped Earth though some recent writers believe that he considered the Earth to be globular. [source: Wikipedia: "Flat Earth"] His other writings make it clear, however, that he considered the Earth to be spherical. He also admitted the possibility of people dwelling at the antipodes [the opposite side of the earth: a notion that presupposes sphericity]. [source: Wikipedia: "Spherical Earth"]

**Paul of Aegina** (c. 625-c. 690) He is considered by some to be the greatest Byzantine surgeon, developed many novel surgical techniques and authored the medical encyclopedia *Medical Compendium in Seven Books*. The book on surgery in particular was the definitive treatise in Europe and the Islamic world for hundreds of years, contained the sum of all Western medical knowledge and was unrivaled in its accuracy and completeness. The sixth book on surgery in particular was referenced in Europe and the Arab world throughout the Middle Ages and is of special interest for surgical history. The whole work in the original Greek was published in Venice in 1528, and another edition appeared in Basel in 1538. [sources: Wikipedia: "Paul of Aegena" and "Science in the Middle Ages"]

**Venerable Bede** (c. 672-735; monk and Doctor of the Church) He understood that the moon affects tides, and that high water was not simultaneous on all the coasts of Britain (recognition of the progressive wave-like character of tides). [source: David Edgar Cartwright, *Tides: A Scientific History* (Cambridge Univ. Press, 2001, pp. 13-14)] *On the Reckoning of Time* (*De temporum ratione*) included an introduction to the traditional ancient and medieval view of the cosmos, including an explanation of how the spherical earth influenced the changing length of daylight ("the roundness of the Earth, for not without reason is it called 'the orb of the world' on the pages of Holy Scripture and of ordinary literature." – section 32), of how the seasonal motion of the Sun and Moon influenced the changing appearance of the New Moon at evening twilight, and a quantitative relation between the changes of the Tides at a given place and the daily motion of the moon. [source: Wikipedia bio and Wikipedia: "Flat Earth"] Bede was lucid about earth's sphericity, writing "We call the earth a globe, not as if the shape of a sphere were expressed in the diversity of plains and mountains, but because, if all things are included in the outline, the earth's circumference will represent the figure of a perfect globe. . . . For truly it is an orb placed in the center of the universe; in its width it is like a circle, and not circular like a shield but rather like a ball, and it extends from its center with perfect roundness on all sides." [source: Wikipedia: "Spherical Earth"]

**Saint Boniface** (c. 672-754) He argued against life in the antipodes of the earth (the other side of the earth). In so doing, he presupposed that the earth was a sphere. [source: White, vol. 1, 104]

**Vergilius of Salzburg** (aka Fergal) (c. 700-784; bishop) He was an early astronomer and believed in a spherical earth. [source]

**Charlemagne** (c. 742-814; Roman emperor)

> Charlemagne . . . and his great minister, Alcuin [c. 740-804], not only promoted medical studies n the schools they founded, but also made provision for the establishment of botanic gardens in which those herbs were especially cultivated which were supposed to have healing virtues.
>
> (from White, vol. II, 34)

**Theodulf of Orléans** (c. 750 to 760-821; bishop) He believed in a spherical earth. [source] A recent doctoral dissertation regarding medieval concepts of the sphericity of the Earth noted that no significant cosmographer since the eighth century has questioned the sphericity of the earth. [source: Wikipedia: "Flat Earth"]

**Dicuil** (late 8th c.; monk) He was an early astronomer and believed in a spherical earth. [source]

**Rabanus Maurus** (c. 780-856; archbishop) He believed in a spherical earth. [source] Isidore's disc-shaped analogy continued to be used through the Middle Ages by authors clearly favouring a spherical Earth, e.g. the 9th century bishop Rabanus Maurus who compared the habitable part of the northern hemisphere (Aristotle's northern temperate clime) with a wheel, imagined as a slice of the whole sphere. [source: Wikipedia: "Spherical Earth"]

**Hunayn ibn Ishaq** (also Hunain or Hunein; 809-873) [Nestorian] His monumental developments on the eye can be traced back to his innovative book, *Ten Treatises on Ophthalmology*: the first systematic book in this field. He explained in minute details about the eye, its diseases and their symptoms and treatments, and its anatomy – all possible by his extensive research and observations. For example, ibn Ishaq taught what cysts and tumors are and the swelling they cause, how to treat various corneal ulcers through surgery, and the therapy involved in repairing cataracts. [source: Wikipedia bio]

**Johannes Scotus Eriugena** (c. 815-c. 877) [universalist] He believed in a spherical earth. [source]

**Remigius of Auxerre** (c. 841-908; Benedictine monk) He believed in a spherical earth. [source]

**Alfred the Great** (849-899; Anglo-Saxon king). He believed in a spherical earth. [source]

**Pope Sylvester II** (born Gerbert d'Aurillac) (c. 940-1003) As a scientist, he was said to be far ahead of his time. Gerbert wrote a series of works dealing with matters of the quadrivium (arithmetic, geometry, astronomy, music), which he taught using the basis of the trivium (grammar, logic, and rhetoric). It is believed that he introduced the use of Arabic figures into Western Europe, and invented the pendulum clock. In Rheims, he constructed a hydraulic-powered organ with brass pipes that excelled all previously known instruments, where the air had to be pumped manually. Gerbert reintroduced the astronomical armillary sphere to Latin Europe via Al-Andalus in the late 10th century as a visual aid for teaching mathematics and astronomy in the classroom; also the abacus. The polar circle on Gerbert's sphere was located at 26 degrees, just several degrees off from the actual 23° 28'. His positioning of the Tropic of Cancer was nearly exact, while his positioning of the equator was exactly correct. [bio sources: Wikipedia, *Catholic Encyclopedia*]

**Notker Labeo** (c. 950-1022; Benedictine monk) He believed in a spherical earth. [source]

## Chapter Three

## 59 Catholic Medieval and/or Scholastic Proto-Scientists from 1000 to 1500 A. D.

**Blessed Hermann of Reichenau** (1013–1054) He wrote several works on geometry and arithmetics and astronomical treatises (including instructions for the construction of an astrolabe, at the time a very novel device in Western Europe). He was disabled, having only limited movement and limited ability to speak. Despite these disabilities he was a key figure in the transmission of Arabic mathematics, astronomy and scientific instruments from Arabic sources into central Europe. He was among the earliest Christian scholars to estimate the circumference of Earth with Eratosthenes' method. [source: Wikipedia bio] He believed in a spherical earth. [source]

**Adelard of Bath** (c. 1080-c. 1152) He was one of the first to introduce the Indian number system to Europe. Adelard also displays original thought of a scientific bent, raising the question of the shape of the Earth (he believed it to be round) and the question of how it remains stationary in space, and also the interesting question of how far a rock would fall if a hole were drilled through the earth and a rock dropped in it, see center of gravity. He theorized that matter could not be destroyed (see Law of conservation of matter) and was also interested in the question of why water experiences difficulty flowing out of a container that has been turned upside down, see atmospheric pressure and vacuum. [source: Wikipedia bio] He contributed the first full Latin translation of Euclid's *Elements* and introduced trigonometry to Europe as transmitted through Arabic astronomical tables. The *Questions Naturales* covers subjects such as plants and animals, the four elements, the hydrological cycle, weather, and astronomy. [source: Scearce]

**Gerard of Cremona** (1114-1187) He translated about 75 books from Arabic into Latin, including works on dialectic, geometry, philosophy, physics, and several other sciences. His activity as a translator helped bring the world of Arabian learning within the reach of the scholars of Latin Christendom and prepared the way for the Scholasticism of the thirteenth century. In this work Gerard was a pioneer. [source: *Catholic Encyclopedia* bio]

**Honorius Augustodunensis** (or Autun) (d. c. 1151; monk) He believed in a spherical earth [source] His Elucidarium (c. 1120) explicitly refers to a spherical earth. [source: Wikipedia: "Flat Earth"] He had a proper appreciation of the value of concrete knowledge. Consequently, he devoted much space in philosophy to the description of the actual world. He thus marks one of the first epochs in the history of the relation between speculative and positive teaching in the Middle Ages. [source: *Catholic Encyclopedia* bio]

**Marius** (fl. 1160) He composed *On the Elements*:

> It is a most remarkable work, employing experiments in a sophisticated if not quite rigorous way, marking a significant advance in the theory of matter, studying with great subtlety the nature of a compound, utilizing a quantitative table to explain how the great variety of the world could arise from just four elements, eschewing magic, and exhibiting a thoroughgoing naturalism in its attitude towards the physical world. This treatise throws much new light on the nature and quality of twelfth-century science and forces a rethinking of the standard accounts of the history of chemistry in the Middle Ages. . . .

Marius explains the formation of stones and metals. Noting that a goldsmith's pot becomes transparent and is changed to glass when subjected to great heat, he concludes that all stones are formed this way in nature by heat enclosed in the interior of the earth....

*De Elementis* . . . is completely naturalistic and materialistic and free of any magical or animistic notions. It therefore represents a kind of early Latin chemistry whose existence has not previously been taken into account in histories of chemistry....

The world of nature obeyed laws, and these laws were accessible to human reason. His presentation of this world does not consist of a string of *ad hoc* explanations, but rather of a carefully worked out, internally consistent, naturalistic scheme, based on a careful and accurate observation of nature and handled with considerable dialectical skill.

We conclude then that at least a part of twelfth-century science was . . . much more than a revival of Antiquity, important though this was. It was bold, original, imaginative and daring. At its best it was rigorous in its arguments and precise in its observations.

(*On the Elements* [Marius], translation and introduction by Richard C. Dales, Univ. of California Press, 1977, pp. 1, 14, 35-36)

**Adam of Bremen** (2nd half of 11th c.) He believed in a spherical earth [source]

**William of Conches** (c. 1090-after 1154) His discussion of meteorology includes a description of air becoming less dense and colder as the altitude increases, and he attempts to explain the circulation of the air in connection with the circulation of the oceans. [source: Wikipedia bio]

**Blessed Hildegard von Bingen** (1098-1179; Benedictine abbess) She believed in a spherical earth [source] and depicted it several times in her work *Liber Divinorum Operum*. She wrote botanical and medicinal texts: *Physica,* on the natural sciences, and *Causae et Curae*. In both texts Hildegard describes the natural world around her, including the cosmos, animals, plants, stones, and minerals. She combined these elements with a theological notion ultimately derived from Genesis: all things put on earth are for the use of humans. [source: Wikipedia bio]

**Rogerius** (c. 1140-c. 1195) He wrote a work on medicine entitled *Practica Chirurgiae* ("The Practice of Surgery"): the first medieval text on surgery to dominate its field in Europe. It laid the foundation for the species of the occidental surgical manuals, influencing them up to modern times. The work, arranged anatomically and presented according to a pathologic-traumatological systematization, includes a brief recommended treatment for each affliction. Rogerius was an independent observer and was the first to use the term lupus to describe the classic malar rash. He recommended a dressing of egg-albumen for wounds of the neck, and did not believe that nerves, when severed, could be regenerated. [source: Wikipedia bio]

**Gerald of Wales** (1147-1220) He understood the basic parameters of the behavior of tides, based on the influence of the moon. He correctly noted that the Atlantic had larger tides than the Mediterranean Sea because of the free course of the tides in the much larger ocean. [source: David Edgar Cartwright, *Tides: A Scientific History* (Cambridge Univ. Press, 2001, pp. 14-15)]

**Robert Grosseteste** (c. 1175–1253; bishop) From about 1220 to 1235 he wrote a host of scientific treatises including:

- *De sphera*. An introductory text on astronomy.
- *De luce*. On the "metaphysics of light." (which is the most original work of cosmogony in the Latin West)
- *De accessu et recessu maris*. On tides and tidal movements. (although some scholars dispute his authorship)

- *De lineis, angulis et figuris*. Mathematical reasoning in the natural sciences.
- *De iride*. On the rainbow. [includes pioneering work on optics]

Grosseteste laid out the framework for the proper methods of science. His work is seen as instrumental in the history of the development of the Western scientific tradition. Grosseteste was the first of the Scholastics to fully understand Aristotle's vision of the dual path of scientific reasoning: generalizing from particular observations into a universal law, and then back again from universal laws to prediction of particulars. Grosseteste called this "resolution and composition". So, for example, looking at the particulars of the moon, it is possible to arrive at universal laws about nature. And conversely once these universal laws are understood, it is possible to make predictions and observations about other objects besides the moon. Further, Grosseteste said that both paths should be verified through experimentation in order to verify the principles. . . . Grosseteste gave a "special importance to mathematics in attempting to provide scientific explanations of the physical world" . . . He saw a key role for geometry in the explanation of natural phenomena. [sources: Wikipedia bio, *Stanford Encyclopedia of Philosophy* bio] "[A]s Grosseteste grew older, he developed increasing reservations about astrology, and in one scientific work after another he gradually abandoned most of its teachings. . . it was bad science, pretending to knowledge it could not have . . ." [source: Dales (in bibliography), pp. 152] "Grosseteste appears to have been the first medieval writer to recognize and deal with the two fundamental methodological problems of induction and experimental verification and falsification which arose when the Greek conception of geometrical demonstration was applied to the world of experience. He appears to have been the first to set out a systematic and coherent theory of experimental investigation and rational explanation by which the Greek geometrical method was turned into modern experimental science." [source: A. C. Crombie, *Robert Grosseteste and the Origins of Experimental Science 1100-1700*; cited in Dales, 172]

**Vincent of Beauvais** (c. 1190-c. 1264; Dominican friar) His *Speculum Naturale* ("Mirror of Nature"), divided into thirty-two books and 3,718 chapters, is a summary of all the science and natural history known to western Europe towards the middle of the 13th century. He accepted the sphericity of the earth. [sources: Wikipedia bio; White, Vol. I, p. 106]

**Gautier de Metz** (mid-13th c.; priest) He believed in a spherical earth [source]

**Bernard of Verdun** (2nd half of 13th c.; Franciscan friar) His most significant work was the Treatise on the Whole of Astronomy (*Tractatus super totam astrologiam*), in which he defended Ptolemy's theory of epicycles and eccentrics and maintained that it was consistent with Aristotle's physics. [source: Wikipedia bio]

**St. Albert the Great (or, Albertus Magnus)** (c. 1193-1280; bishop and Doctor of the Church) He accepted the sphericity of the earth [source: Hannam: "The Myth of the Flat Earth"; White, vol. I, p. 97; additional source] Albertus' writings collected in 1899 went to thirty-eight volumes. These displayed his prolific habits and literally encyclopedic knowledge of topics such as botany, astronomy, mineralogy, chemistry, zoology, physiology and others; all of which were the result of logic and observation. Albertus' knowledge of physical science was considerable and for the age remarkably accurate. His industry in every department was great. He is credited with the discovery of the element arsenic. [source: Wikipedia bio] He made contributions to logic, psychology, metaphysics, meteorology, mineralogy, and zoology. [source: *Stanford Encyclopedia of Philosophy* bio] He became famous for his vast knowledge and for his defence of the pacific coexistence between science and religion. Albert was an essential figure in introducing Greek and Islamic science into the medieval universities, although not without hesitation with regard to particular Aristotelian theses. In one of his most famous sayings he asserted: "Science does not consist in ratifying what others say, but of searching for the causes of phenomena." [source: Wikipedia: "Science in the Middle Ages"]

**Johannes de Sacrobosco** (c. 1195-c. 1256) His work, *Tractatus de Sphaera*, the most influential astronomy textbook of the 13th century and required reading by students in all Western European universities, described the earth as a sphere. Its popularity exposes the nineteenth-century opinion that medieval scholars thought the earth was flat as a fabrication. His *Algorismus* was the first text to introduce Hindu-Arabic numerals and procedures into the European university curriculum. [source: Wikipedia bio]

**Bartholomew of England** (c. 1203-1272; Franciscan friar and bishop) He studied under Robert Grosseteste and was the author of *On the Properties of Things (De proprietatibus rerum)*, an early forerunner of the encyclopedia. It has sections on physiology, medicine, the universe and celestial bodies, time, form and matter (elements), air and its forms, water and its forms, earth and its forms including geography, gems, minerals and metals, animals, and color, odor, taste and liquids. It was the first to make readily available the views of Greek, Jewish, and Arabic scholars on medical and scientific subjects. [source: Wikipedia bio]

**Theodoric Borgognoni** (1205-1298; Dominican friar and bishop) His major medical work is the *Cyrurgia*, a systematic four-volume treatise covering all aspects of surgery. He insisted that the practice of encouraging the development of pus in wounds, handed down from Galen and from Arabic medicine be replaced by a more antiseptic approach, with the wound being cleaned and then sutured to promote healing. Bandages were to be pre-soaked in wine as a form of disinfectant. He also promoted the use of anesthetics in surgery. A sponge soaked in a dissolved solution of opium, mandrake, hemlock, mulberry juice, ivy and other substances was held beneath the patients nose to induce unconsciousness. Borgognoni's test for the diagnosis of shoulder dislocation, namely the ability to touch the opposite ear or shoulder with the hand of the affected arm, has remained in use into modern times. [source: Wikipedia bio]

**William of Saliceto** (1210-1277) In 1275 he wrote *Chirurgia* which promoted the use of a surgical knife over cauterizing. He also was the author of *Summa conservationis et curationis* on hygiene and therapy and gave lectures on the importance of regular bathing for infants, and special care for the hygiene of pregnant women. [source: Wikipedia bio]

**Gerard of Brussels** (early 13th c.) Known primarily for his Latin book *Liber de motu* (or *On Motion*), which was a pioneering study in kinematics. His chief contribution was in moving away from Greek mathematics and closer to the notion of "a ratio of two unlike quantities such as distance and time", which is how modern physics defines velocity. [source: Wikipedia bio]

**Roger Bacon** (c. 1214-1294; Franciscan friar) He is sometimes credited as one of the earliest European advocates of the modern scientific method and he strongly championed experimental study. He urged theologians to study all sciences closely, and to add them to the normal university curriculum. His *Opus Majus* contains treatments of mathematics and optics, the manufacture of gunpowder, the positions and sizes of the celestial bodies, and anticipates later inventions such as microscopes, telescopes, spectacles, hydraulics and steam ships. The study of optics in part five of *Opus Majus* seems to draw on the works of the Muslim scientists, Alkindus (al-Kindi) and Alhazen (Ibn al-Haytham), including a discussion of the physiology of eyesight, the anatomy of the eye and the brain, and considers light, distance, position, and size, direct vision, reflected vision, and refraction, mirrors and lenses. Bacon predicted the invention of the submarine, automobile, and airplane. Bacon's writings consider Newtonian metrical frameworks for space, then reject these for something which reads remarkably like Einsteinian Relativity. [source: Wikipedia bio] Natural causation occurs "naturally" according to regular processes or laws of nature. There is no "spiritual being" in the medium as was commonly taught by other Scholastic philosophers. No, for Bacon, universal causation is corporeal and material, and matter itself in not just pure potentiality but is rather something *positive in itself.* He refers to the "laws of reflection and refraction" as *leges communes nature.* For Bacon in his account of nature in *Communia naturalium* and the later works in general, a general law of nature governs universal force. This universal law of nature is imposed on a world of Aristotelian natures. This notion would have a significant future in experimental science. Starting from Aristotle's account of *empeiria (experience)*, Bacon argues that logical argument alone, even when it originates from experience, is not sufficient for the "verification of things." Even arguments that have their origins in experience will need to be verified by means of an intuition of the things in the world. He distinguishes "natural scientific argument" from moral and religious mystical intuition. He calculated the measured value of 42 degrees for the maximum elevation of the rainbow. This was probably done with an astrolabe, and in this, Bacon advocates the skillful mathematical use of instruments for an experimental science. There is much evidence that Bacon himself did mathematical work and experiments with visual phenomena such as pinhole images and the measurement of the visual field. [source: *Stanford Encyclopedia of Philosophy* bio]

**Pierre de Maricourt** (or, Petrus Peregrinus) (fl. c. 1269) He was the first to give an account of magnetic Polarity and methods for determining the poles of a magnet. He may also have been the first to apply the term *Polus* to magnetic pole). He was also aware that Polaris, the pole star, does not rest at the celestial north pole, but revolves around it. From this knowledge Peregrinus concluded that the poles of a magnet, or magnetized needle, always point directly to the celestial poles rather than to the pole star, as commonly believed. His *Epistola* was the first extant treatise devoted exclusively to magnetism, creating the first science of magnetism. He formulated rules for the enabled him to enunciate rules for attraction and repulsion, all of which would today form the basis of an introductory lesson on magnetism. [source: Encyclopedia.com bio] The permanent magnetization of iron, the properties of the magnetic poles, the direction of the Earth's action exerted on these poles or of their action on one another, are all found very accurately described in [his] treatise written in 1269: a model of the art of logical sequence between experiment and deduction. [source: Wikipedia: "History of Physics"]

**Bertold of Regensburg** (c. 1220-1272; Franciscan monk) He believed in a spherical earth [source] The fact that he used the spherical Earth as a sermonic illustration shows that he could assume this knowledge among his congregation. The sermon was held in the vernacular German, and thus was not intended for a learned audience. [source: Wikipedia: "Flat Earth"]

**William of Auvergne** (d. 1249) He compared the attraction of the tides to the moon to that of a magnet for iron. [source: *Catholic Encyclopedia*: "History of Physics"]

**Jordanus de Nemore** (mid-13th c.) The medieval "science of weights" (i.e., mechanics) owes much of its importance to the work of Jordanus. In the *Elementa super demonstrationem ponderum*, he introduces the concept of "positional gravity" and the use of component forces. He proves the law of the lever by means of the principle of work. The *De ratione ponderis* also proves the conditions of equilibrium of unequal weights on planes inclined at different angles – long before Galileo. Jordanus' *De numeris datis* was the first treatise in advanced algebra composed in Western Europe, building on elementary algebra provided in twelfth-century translations from Arabic sources. It anticipates by 350 years the introduction of algebraic analysis by François Viète into Renaissance mathematics. He also analyzed the mathematics of stereographic projection. [source: Wikipedia bio]

**St. Thomas Aquinas** (1225-1274; Dominican friar) His greatest contribution to the scientific development of the period was having been mostly responsible for the incorporation of Aristotelianism into the Scholastic tradition, and in particular his

*Commentary on Aristotle's Physics* was responsible for developing one of the most important innovations in the history of physics, namely the notion of the inertial resistant mass of all bodies universally, subsequently further developed by Kepler and Newton in the 17th century. [source: Wikipedia: "Science in the Middle Ages"] He accepted the sphericity of the earth [source: Hannam: "The Myth of the Flat Earth"; White, vol. I, p. 97; additional source] John F. McCarthy explains how St. Thomas, following St. Augustine, arguably espoused something similar to theistic evolution (or at least its possibility):

> A final question regards the rise of new forms of corporeal existence. St. Augustine understands the implanting of the forms of the various creatures described in Genesis 1 to have taken place simultaneously with the act of creation in the beginning, although he also distinguishes between the implantation of some forms in the act of existence and the implantation of other forms causally in the potency of the matter. (1) St. Thomas sees no contradiction in this interpretation, but he also points out that forms are not "implanted" in the sense that they first exist outside of their subject and then are instilled within. What actually happens is that both this matter and this form come into existence simultaneously as this individual thing. (2) But the question remains as to how higher forms can exist in the potency of the matter of lower forms. St. Augustine sees plants as having existed originally in the potency of the earth, and he goes on to say that, over the years down to the present, God "plants" living things, as He planted the verdure of Paradise, in His ordinary governance of all things. (3) Nevertheless, what needs to be clarified is the sense in which God "plants" what was already created causally from the beginning. St. Thomas cites Aristotle to the effect that for the generation of some vegetation all that is needed is the power of the physical heaven in place of the father and the power of the earth in place of the mother, (4) . . .

While St. Thomas affirms that new bodily forms arise by the interaction of bodies upon one another, he also requires the intervention of God for the first production of things.

> In the first establishment (*institutione*) of things, the active principle was the Word of God, which from elemental matter produced animals either in act according to some of the Fathers or virtually according to Augustine. Not that water or earth has in itself the power to produce all of the animals, as Avicenna claimed, but the fact that animals can be produced from elemental matter by the power of seed or of the heavenly bodies comes from a power initially given to the elements. (5)

It is easily within the power of God to have caused the mountains and the oceans to take shape in a few hours . . . to have spread out the universe in an equally short time, and to have created streams of light from the most distant galaxies just on the point of reaching the earth. But it is by no means necessary to believe that God did this, and no one should insist that the text of Genesis demands such a reading. St. Augustine (6) and St. Thomas (7) both point out that it would not have been contrary to divine wisdom for God to have performed the work of creation according to a pattern that natural processes would afterwards imitate, and it is known today that natural processes tend to follow a developmental pattern. St. Augustine and St. Thomas also warn against unnecessarily defending readings of the Scripture which go against what natural science and experience seem to indicate, as is taken to be the case with the 24-hour interpretation of the six days of creation. The text of Genesis 1 is open to the interpretation of the six days of creation as six undefined periods of time which are called days because they are sub-divided into a time of darkness followed by a time of light . . .

St. Augustine explains, or rather hypothesizes, that in one sense the entire creation took place in an instant, and, therefore, there was no problem of plant life's being said to have existed before the sun. But, in affirming this, St. Augustine also distinguishes between what was created in actual being and what was created potentially in the packages of powers ("seminal reasons") with which God endowed elemental matter in the first instant of creation. In the view of St. Augustine, primal matter developed upward after the first moment of its creation because of the plan of development that God had instilled in it and because of certain formative interventions that God continued to make even after the first six days of creation. We must admit that Augustine does not attempt clearly to determine how much upward "development" was already included in the original instant of creation and how much came after that instant. (8) . . .

. . . the idea of a long period of development does not in itself conflict with either the letter or the spirit of the Scriptures; it simply illustrates the transcendence and the eternity of God, for Whom a thousand million years is not even one instant in our psychological time-experience.

Notes

1) Aquinas, *Summa Theologica* (S. Th.) I, q. 69, art. 2, corp. Cf. LT 47, pp. 5-7.
2) *S. Th.* I, q. 65, art. 4, corp.
3) Augustine, *De Gen. ad litt.*, V, 4.
4) Aquinas, II Sent., dist. 14, q. 1, art. 5, ad 6.
5) *S. Th.* I, q. 71, art. 1, ad 1.
6) Cf. Augustine, *De Gen. ad litt.*, II, 15.
7) Cf. Aquinas, *S. Th.*, I, q. 74, art. 2, ad 4.
8) Cf. Augustine, *De Gen. ad litt.*, V, 4.

("A Neo-Patristic Return to the First Four Days of Creation," Part VI)

He believed in a spherical Earth; and he even took for granted his readers also knew the Earth is round. In his Summa Theologica, [1st part of the 2nd part, Q. 54] he wrote, "The physicist proves the earth to be round by one means, the astronomer by another: for the latter proves this by means of mathematics, e.g. by the shapes of eclipses, or something of the sort; while the former proves it by means of physics, e.g. by the movement of heavy bodies towards the center, and so forth." [source: Wikipedia: "Flat Earth"]

> As an Aristotelian natural philosopher and a professional theologian, one may appropriately inquire how Thomas related natural philosophy and theology, the medieval equivalent of the relations between science and religion. Thomas followed in the path of his teacher, Albert the Great, and generally refrained from introducing theological ideas into his treatises on natural philosophy, whereas he did not hesitate to introduce natural philosophy to elucidate his theological discussions. As a theologian doing natural philosophy, Thomas could easily have resorted to theological appeals and arguments in his natural philosophy, but he did not think it appropriate to do so. As he explained in a reply to one of forty-three questions sent to him by the master general of the Dominican order, "I don't see what one's interpretation of the text of Aristotle has to do with the teaching of the faith." Thomas refused to Christianize Aristotle's natural philosophy and to confuse natural philosophy with theology. In this, Thomas followed the practice of most medieval theologians and natural philosophers.
>
> (*Encyclopedia of Science and Religion* [online book], "Thomas Aquinas")

Aquinas seems to have left open a distinct possibility of heliocentrism or some explanation other than geocentrism:

Reason may be employed in two ways to establish a point: firstly, for the purpose of furnishing sufficient proof of some principle, as in natural science, where sufficient proof can be brought to show that the movement of the heavens is always of uniform velocity. Reason is employed in another way, not as furnishing a sufficient proof of a principle, but as confirming an already established principle, by showing the congruity of its results, as in astrology the theory of eccentrics and epicycles is considered as established, because thereby the sensible appearances of the heavenly movements can be explained; not, however, as if this proof were sufficient, forasmuch as some other theory might explain them.

(*Summa Theologiae*, First Part, Q. 32, Article 1. Whether the trinity of the divine persons can be known by natural reason?, Reply to Objection 2)

He also rejected astrology (something that Galileo, Tycho Brahe, and Kepler were still enthralled with some 300 or more years later). Some have argued that he espoused it in his *Summa Theologica*, in a section called, Whether divination by the stars is unlawful? (Second part of Second part, question 95). It appears, however, that Aquinas was simply accepting aspects of astrology that had some semblance of scientific value in them: aspects of star-watching that were far closer to astronomy than to the occult. Science was not as fully developed in his time, so we would expect to see some such confusion (and partially it was a matter of semantics). For example, if astrologers predicted a solar eclipse, then obviously they had made some observation that was scientific, in that it recognized observable patterns in the sky ("it is evident that those things which happen of necessity can be foreknown by this mean, even so astrologers forecast a future eclipse."). Note that the three "objections" are not the opinion of St. Thomas. He goes on to dispute the fundamental thesis of astrology: that the stars affect human behavior and decisions, etc.:

> In the second place, acts of the free-will, which is the faculty of will and reason, escape the causality of heavenly bodies. For the intellect or reason is not a body, nor the act of a bodily organ, and consequently neither is the will, since it is in the reason, as the Philosopher shows (*De Anima* iii, 4,9). Now no body can make an impression on an incorporeal body. Wherefore it is impossible for heavenly bodies to make a direct impression on the intellect and will . . .

He denies the false, occultic part of astrology (which is, of course, the great bulk of it):

> Accordingly if anyone take observation of the stars in order to foreknow casual or fortuitous future events, or to know with certitude future human actions, his conduct is based on a false and vain opinion; and so the operation of the demon introduces itself therein, wherefore it will be a superstitious and unlawful divination.

But he accepts that which simply operates on the same principles as science:

> On the other hand if one were to apply the observation of the stars in order to foreknow those future things that are caused by heavenly bodies, for instance, drought or rain and so forth, it will be neither an unlawful nor a superstitious divination.

All that St. Thomas Aquinas really grants here is some influence of the stars and planets on humans insofar as this is explained in terms of physical causation. That doesn't involve the occult. We know, for instance, of the influence of the moon on tides. The theory of gravity involves relationships between physical bodies in space. We are pulled to the earth: so the earth itself "influences" our bodies in that way. There seems to be some rrelationship with lunar cycles and psychologically disturbed people (the etymological background of the word *lunatic*). He acknowledges that *some* truth can be found almost anywhere, but when all is said and done, he ends up by citing St. Augustine in strong disagreement with astrology:

> Thus a good Christian should beware of astrologers, and of all impious diviners, especially of those who tell the truth, lest his soul become the dupe of the demons and by making a compact of partnership with them enmesh itself in their fellowship.

**Arnaldus de Villa Nova** (1235-1311) He is credited with translating a number of medical texts from Arabic, including works by Ibn Sina (Avicenna), Qusta ibn Luqa (Costa ben Luca), and Galen. He is also the reputed author of various medical works, including *Breviarium Practicae*. He discovered carbon monoxide and pure alcohol. [source: Wikipedia bio]

**Richard of Middletown** (late 13th c.) He knew and studied the phenomena of hypnotism, and left the results of his investigations in his *Quodlibeta* where he treats of what would now be termed auto-suggestion. [source: Catholic Encyclopedia bio] Richard of Middletown (about 1280) and, after him, many masters at Paris and Oxford admitted that the laws of nature are certainly opposed to the production of empty space, but that the realization of such a space is not, in itself, contrary to reason; thus, without any absurdity, one could argue on vacuum and on motion in a vacuum. He and Duns Scotus (about 1275-1308) began to formulate hypotheses to the effect that the heavenly bodies were animated by other [non-rotary] motions, and the entire school of Paris adopted the same opinion. Soon, however, the Earth's motion was taught in the School of Paris, not as a possibility, but as a reality. He believed that God could create other worlds similar to ours. [source: *Catholic Encyclopedia*: "History of Physics"]

**Henry of Ghent** (d. 1293) He believed that God could create other worlds similar to ours. [source: Wikipedia: "History of Physics"]

**Gregory Choniades** (d. c. 1302) He translated a number of Arabic and Persian works on mathematics and astronomy and played an important role in transmitting several innovations from the Islamic world to Europe. These include the introduction of the universal latitude-independent astrolabe to Europe and a Greek description of the Tusi-couple, which would later have an influence on Copernican heliocentrism. [source: Wikipedia bio]

**Theodoric [or Thierry or Dietrich] of Freiberg** (c. 1250–c. 1310; Dominican monk) Drawing from his two earlier works on light and colour, he wrote *De iride et radialibus impressionibus* (*On the Rainbow and the impressions created by irradiance*, c. 1304-1311), relying on geometry, experiment, falsification and other methods. Among other properties he explained in detail:

- the colors of the primary and secondary rainbows
- the positions of the primary and secondary rainbows
- the path of sunlight within a drop: lightbeams are refracted when entering the atmospheric droplets, then reflected inside the droplets and finally refracted again when leaving them.
- the formation of the rainbow: he explains the role of the individual drops in creating the rainbow
- the phenomenon of color reversal in the secondary rainbow

Using spherical flasks and glass globes filled with water, Freiberg was able to simulate the water droplets during rainfall. Still in its early stages, experimental instrumentation would later expand to be used primarily for making measurements, extending the human senses and creating and isolated environment for the experimenter. During his experimentation with these glass globes, Freiberg was correct in asserting that the colors formed in interaction with the water droplets. [Wikipedia bio] These studies were models of the art of logically combining experiments. His scientific treatises on light (*De luce*), on color (*De coloribus*), and on the rainbow (*De iride*) contributed greatly to the development of optics and were composed in the scientific spirit of Albertus Magnus. [source: _Stanford Encyclopedia of Philosophy_ bio]

**Henri de Mondeville** (c. 1260-1316) The "Father of French Surgery" and author of _Cyrurgia_ (Surgery): written in 1312: the first medieval treatise on the subject. [source: Wikipedia bio]

**Meister Eckhart** (c. 1260-c. 1327; Dominican friar) He believed in a spherical earth [source]

**Dante Alighieri** (c. 1265-1321) His _Divine Comedy_ portrays the earth as a sphere, discussing implications such as the different stars visible in the southern hemisphere, the altered position of the sun, and the various timezones of the earth. [source: Wikipedia: "Flat Earth"]

**Mondino de Luzzi** (c. 1270-1326) He is often credited as the "restorer of anatomy" because he made seminal contributions to the field by reintroducing the practice of public dissection of human cadavers and writing the first modern anatomical text: _Anathomia corporis humani_. He describes the closure of an incised intestinal wound by having large ants bite on its edges and then cutting off their heads, which one scholar interprets as an anticipation of the use of staples in surgery. For three centuries, the statutes of many medical schools required lecturers on anatomy to use _Anathomia_ as their textbook. [source: Wikipedia bio]

**Joannes Zacharias Actuarius** (c. 1275-c. 1328) He wrote _De Urinis_: a treatise on urine in seven books. Actuarius treated of this subject fully and distinctly, and, though he goes upon the plan which Theophilus Protospatharius [7th c.] had marked out, yet he has added a great deal of original matter. It is the most complete and systematic work on the subject that remains from antiquity, so much so that, till the chemical improvements of the 19th century, he had left hardly anything new to be said by the moderns. [source: Wikipedia bio]

**Walter Burley** (c. 1275-1344) He believed that God could create other worlds similar to ours. [source: *Catholic Encyclopedia*: "History of Physics"]

**Guidelines or Condemnations for the University of Paris - 1277** An event of note in the thirteenth century was a promulgation of 219 propositions related to Greek science, primarily as guidelines for the University of Paris. This was initiated by the Pope and dealt with most of the matters that had exercised the Christian thinkers of the previous twelve centuries. The list included the following: rejection of the eternity of the world and of the cyclic recurrence of its life every 36,000 years; the natural world was uniform in its constitution and laws, and stood in a contingent relation to its Creator; rejection of the heavenly bodies being animated and incorruptible, and of the influence of the stars upon human lives; and acceptance of the possibility of linear motion for the heavenly bodies, instead of the circular movement obligatory in Greek science. [source: Tripp] From at least 1280 onward, many masters at Paris and Oxford admitted that the laws of nature are certainly opposed to the production of empty space, but that the realisation of such a space is not, in itself, contrary to reason. These arguments gave rise to the branch of mechanical science known as dynamics. Historians agree that the condemnations allowed science to consider new possibilities that Aristotle never conceptualized. According to the historian of science Richard Dales, they "seem definitely to have promoted a freer and more imaginative way of doing science" ["The De- Animation of the Heavens in the Middle Ages," *Journal of the History of Ideas* 41 (1980): 531-50; quote from p. 550]. The new philosophy of nature, that emerged from the rise of Skepticism following the Condemnations, contained "the seeds from which modern science could arise in the early seventeenth century." [David C. Lindberg, *Science in the Middle Ages* (1980), p. 111] [source: Wikipedia: "Condemnations of 1210-1277"] [see also: *Stanford Encyclopedia of Philosophy*, "Condemnations of 1277"]

**Francis of Marchia** (c. 1285-c. 1344; Franciscan friar) Marchia displays a great interest in the causal process. Notable is his clear and sharp distinction between natural causation that works necessarily and the contingent causation of human and divine free will. One of the most important innovations of the mature Galileo was the assertion that the celestial and terrestial realms are made of the same fundamental matter and therefore follow the same basic natural laws. Francis of Marchia put forth a hypothesis similar to Galileo's in his commentary on book II of the *Sentences*: that the heavens are not made up of matter so completely different from terrestrial matter that it radically differentiates the supralunar realm from the sublunar one. On the contrary, the basic matter is the same everywhere, and just as Marchia considers the natural world to follow predictable patterns, he also thinks that those patterns are universally applicable. Marchia also most probably had an impact on the development of the theory of inertia. [source: *Stanford Encyclopedia of Philosophy* bio]

**William of Ockham** (or Occam) (c. 1288- c. 1348; Franciscan friar) One important contribution that he made to modern science and modern intellectual culture was through the principle of parsimony in explanation and theory building that came to be known as Occam's Razor. The principle says that one should not multiply entities beyond necessity. He formulates it as: "For nothing ought to be posited without a reason given, unless it is self-evident (literally, known through itself) or known by experience or proved by the authority of Sacred Scripture." [source: Wikipedia bio] "While Scholastic in setting," as David Lindberg writes, it was "thoroughly modern in orientation. Referred to as the *via moderna*, in opposition to the *via antiqua* of the earlier scholastics, it has been seen as a forerunner of a modern age of analysis." [*Science in the Middle Ages* (1980), p. 109] [source: Wikipedia: "Condemnations of 1210-1277"] He believed that God could create other worlds similar to ours. By means of most spirited argumentation, he made known the complete absurdity of the Peripatetic theory of the motion of projectiles. He thought that the matter constituting celestial bodies was of the same nature as that constituting sublunary bodies and that, if the former remained perpetually the same, it was not because they were, by nature, incapable of change and destruction, but simply because the place in which they were contained no agent capable of corrupting them. [source: *Catholic Encyclopedia*: "History of Physics"] Frederick Copleston, the eminent historian of philosophy, observed:

> [I]t is probably true to say that Ockhamist insistence on experience as the basis of our knowledge of existent things favoured the growth of empirical science . . . Ockhamism . . . helped to create an intellectual climate which facilitated and tended to promote scientific research. For by directing men's minds to the facts or empirical data in the acquisition of knowledge it at the same time directed them away from passive acceptance of the opinions of illustrious thinkers of the past. . . .
> 
> One can say then, I think, that the leading figures in the scientific movement of the fourteenth century had in most cases affiliations with the Ockhamist Movement.
> 
> (*A History of Philosophy, Volume 3: Late Medieval and Renaissance Philosophy*, part I [Doubleday Image edition, 1963], 166-167)

In his book, *A Brief History of Time* (Bantam: 1988, p. 55), Stephen Hawking attributes the discovery (by Heisenberg, Schrodinger, and Dirac) of quantum mechanics to Ockham's Razor.

**Gentile da Foligno** (d. 1348) Gentile wrote several widely copied and read texts and commentaries, notably his massive commentary covering all five books of the *Canon of Medicine* by the 11th-century Persian polymath Avicenna, the comprehensive encyclopedia that, in Latin translation, was fundamental to medieval medicine. Gentile's commentary *de urinarum iudiciis* made the first attempt to comprehend the physiology of urine formation: asserting that urine associated with the blood passes "through the porous tubules" of the kidney and is then delivered to the bladder. He connected the relationship between fast pulse rate and urine output and correlated the color of urine with the condition of the heart. For the originality of his thought it has been suggested that he was the first cardionephrologist. [source: Wikipedia bio]

**Thomas Bradwardine** (c. 1290–1349; archbishop) Merton College sheltered a group of dons devoted to natural science, mainly physics, astronomy and mathematics, rivals of the intellectuals at the University of Paris. Bradwardine was one of these Oxford Calculators, studying mechanics with William Heytesbury, Richard Swineshead, and John Dumbleton. The Oxford Calculators distinguished kinematics from dynamics, emphasizing kinematics, and investigating instantaneous velocity. They first formulated the mean speed theorem: a body moving with constant velocity travels the same distance as an accelerated body in the same time if its velocity is half the final speed of the accelerated body. They also demonstrated this theorem—the essence of "The Law of Falling Bodies" — long before Galileo, who is generally credited with it. [source: Wikipedia bio]

**Guy de Chauliac** (c. 1300-1368) He was among the most important physicians of his time, and his ideas dominated surgical thought for over 200 years. He is most famous for his work on surgery, *Chirurgia magna*. In seven volumes, it covers anatomy, bloodletting, cauterization, drugs, anesthetics, wounds, and fractures, ulcers, special diseases, and antidotes. His treatments included the use of plasters. He also wrote *De ruptura*, which describes different types of hernias; and *De subtilianti diaeta*, explaining cataracts and possible treatments for them. [source: Wikipedia bio]

**Richard Kilvington** (c. 1302-1361) It is his analysis of local motion that places Kilvington among the 14th-century pioneers who considered the problem of motion with respect to its causes (*tamquam penes causam*), corresponding to modern dynamics, and with respect to its effects (*tamquam penes effectum*), corresponding to modern kinematics. In his works on the philosophy of nature, he raised many fundamental questions, often solving them in an original and sophisticated manner. [source: *Stanford Encyclopedia of Philosophy* bio]

**William Crathorn** (fl. 1330s) The implications of Crathorn's atomism are truly astonishing. First, every movement boils down to local motion of atoms in the void. Thus, Crathorn affirms that a continuous motion has only one possible speed, which is the greatest speed it could ever reach. In other words, movement is continuous when an atom changes from one atomic place to another contiguous atomic place. The proportion of time and place (i.e., the speed) is always equal to one. [source: *Stanford Encyclopedia of Philosophy* bio]

**Albert of Saxony** (or Helmstadt) (mid-14th c.) A principle was formulated which for three centuries was to play a great role in statics, viz. that every heavy body tends to unite its centre of gravity with the centre of the Earth. When writing his "Questions" on Aristotle's "De Cælo" in 1368, Albert admitted this principle, which he applied to the entire mass of the terrestrial element. The centre of gravity of this mass is constantly inclined to place itself in the centre of the universe, but, within the terrestrial mass, the position of the centre of gravity is incessantly changing. Now, in order to replace this centre of gravity in the centre of the universe, the Earth moves without ceasing; and meanwhile a slow but perpetual exchange is being effected between the continents and the oceans. He ventured so far as to think that these small and incessant motions of the Earth could explain the phenomena of the precession of the equinoxes, and declared that one of his masters, whose name he did not disclose, announced himself in favour of the daily rotation of the Earth, inasmuch as he refuted the arguments that were opposed to this motion. He adopted Buridan's theory of dynamics in its entirety. [source: Wikipedia: "History of Physics"] His treatises contain, in a clear, precise, and concise form, an explanation of numerous ideas which exercised great influence on the development of modern science. He abandoned the old Peripatetic dynamics which ascribed the movement of projectiles to disturbed air. With Buridan he placed the cause of this movement in an impetus put into the projectile by the person who threw it; the part he assigned to this impetus is very like that which we now attribute to living force. He considered that the heavens were not moved by intelligences, but, like projectiles, by the impetus which God gave them when He created them, and saw in the increase of impetus the reason of the acceleration in the fall of a heavy body. He further taught that the velocity of a falling weight increased in proportion either to the space traversed from the beginning of the fall or to the time elapsed, but he did not decide between these two. [source: Catholic Encyclopedia bio]

**Richard Swineshead** (fl. c. 1340-1354) Mathematician, logician, and natural philosopher. He was perhaps the greatest of the Oxford Calculators of Merton College. His magnum opus was a series of treatises known as the *Liber calculationum* ("Book of Calculations"). Gottfried Leibniz wrote in a letter of 1714: "his works are little known, but what I have seen of them seemed to me profound and relevant." [source: Wikipedia bio]

**Jean Buridan** (1300-1358; priest) anticipated the idea of inertia (the idea an object once in motion continues to move in the same direction until it encounters resistance) through his discussion of impetus:

> Also, since the Bible does not state that appropriate intelligences move the celestial bodies, it could be said that it does not appear necessary to posit intelligences of this kind, because it would be answered that God, when He created the world, moved each of the celestial orbs as He pleased, and in moving them He impressed in them impetuses which moved them without His having to move them any more except by the method of general influence whereby He concurs as a co-agent in all things which take place; 'for thus on the seventh day He rested for all work .

. .' [Gen. 2:2] And these impetuses which He impressed in the celestial bodies were not decreased nor corrupted afterwards, because there was not inclination of the celestial bodies for movements. . . .

But because of the resistance which results from the weight of the [waterwheel of the] mill, the impetus would continually diminish until the mill ceased to turn. And perhaps, if the mill should last forever without any diminution or change, and there were no other resistance to corrupt the impetus, the mill would move forever because of its perpetual impetus. [source: Snow]

He rejected the Aristotelian idea [in *De Caelo*] of a cosmos existing from all eternity. Following in the footsteps of John Philoponus and Avicenna, proposed that motion was maintained by some property of the body, imparted when it was set in motion. Buridan named the motion-maintaining property *impetus*. Moreover, he rejected the view that the impetus dissipated spontaneously (this is the big difference between Buridan's theory of impetus and his predecessors), asserting that a body would be arrested by the forces of air resistance and gravity which might be opposing its impetus. Buridan further held that the impetus of a body increased with the speed with which it was set in motion, and with its quantity of matter. Clearly, Buridan's impetus is closely related to the modern concept of momentum. Buridan saw impetus as *causing* the motion of the object. Buridan anticipated Isaac Newton when he wrote:

. . . after leaving the arm of the thrower, the projectile would be moved by an impetus given to it by the thrower and would continue to be moved as long as the impetus remained stronger than the resistance, and would be of infinite duration were it not diminished and corrupted by a contrary force resisting it or by something inclining it to a contrary motion. [source: Wikipedia bio]

Buridan's account of motion is in keeping with his approach to natural science, which is empirical in the sense that it emphasizes the evidentness of appearances, the reliability of a posteriori modes of reasoning, and the application of naturalistic tropes or models of explanation (such as the concept of impetus) to a variety of phenomena. [source: Stanford Encyclopedia of Philosophy bio] With the assistance of these principles concerning impetus, Buridan accounts for the swinging of the pendulum. He likewise analyses the mechanism of impact and rebound and, in this connexion, puts forth very correct views on the deformations and elastic reactions that arise in the contiguous parts of two bodies coming into collision. Nearly all this doctrine of impetus is transformed into a very correct mechanical theory if one is careful to substitute the expression *vis viva* for impetus. [source: Wikipedia: "History of Physics"]

**Konrad of Megenberg** (1309-1374) His best-known and most widely read work is his *Buch der Natur*: a survey of all that was known of natural history at that time and is, besides, the first natural history in the German language. The work has eight chapters, on the nature of man; sky, seven planets, astronomy and meteorology; zoology; ordinary and aromatic trees; plants and vegetables; invaluable and semi-precious stones; ten kinds of metals; water and rivers. [source: Wikipedia bio]

**William Heytesbury** (c. 1313-c. 1373) His work curiously anticipates nineteenth-century mathematical analysis of the continuum, and he is well know for developing the Mean Speed Theorem concerning the distance covered over a period of time under uniform acceleration. His work had some influence on the development of early modern science. [source: Stanford Encyclopedia of Philosophy bio]

**Giovanni di Casali** (d. c. 1375; Franciscan friar) About 1346 he wrote a treatise *On the Velocity of the Motion of Alteration*. In it he presented a graphical analysis of the motion of accelerated bodies. His teachings in mathematical physics is believed to have influenced the similar ideas presented over two centuries later by Galileo. [source: Wikipedia bio]

**Nicholas Oresme** (c. 1323-1382; bishop) Oresme conceived the idea of employing what we should now call rectangular co-ordinates . . . and thus forestalls Descartes in the invention of analytical geometry. . . . In opposition to the Aristotelean theory of weight, according to which the natural location of heavy bodies is the centre of the world, and that of light bodies the concavity of the moon's orb, he proposes the following: The elements tend to dispose themselves in such manner that, from the centre to the periphery their specific weight diminishes by degrees. He thinks that a similar rule may exist in worlds other than this. This is the doctrine later substituted for the Aristotelean by Copernicus and his followers . . . But Oresme had a much stronger claim to be regarded as the precursor of Copernicus when one considers what he says of the diurnal motion of the earth, . . . He begins by establishing that no experiment can decide whether the heavens move form east to west or the earth from west to east; for sensible experience can never establish more than one relative motion. He then shows that the reasons proposed by the physics of Aristotle against the movement of the earth are not valid . . . [source: *Catholic Encyclopedia*: "Nicole Oresme"] He wrote influential works on mathematics, physics, and astronomy. In his *Livre du ciel et du monde* Oresme discussed a range of evidence for and against the daily rotation of the Earth on its axis. From astronomical considerations, he maintained that if the Earth were moving and not the celestial spheres, all the movements that we see in the heavens that are computed by the astronomers would appear exactly the same as if the spheres were rotating around the Earth. He rejected the physical argument that if the Earth were moving the air would be left behind causing a great wind from east to west. In his view the Earth, Water, and Air would all share the same motion. As to the scriptural passage that speaks of the motion of the sun, he concludes that "this passage conforms to the customary usage of popular speech" and is not to be taken literally. He also noted that it would be more economical for the small Earth to rotate on its axis than the immense sphere of the stars. [source: Wikipedia bio] His work provided some basis for the development of modern mathematics and science. Oresme brilliantly argues against any proof of the Aristotelian theory of a stationary Earth and a rotating sphere of the fixed stars and showed the possibility of a daily axial rotation of the Earth. He was a determined opponent of astrology, which he attacked on religious and scientific grounds. He states – more than 300 years before Robert Hooke (1635–1703) and Newton – that atmospheric refraction occurs along a curve and proposes to approximate the curved path of a ray of light in a medium of uniformly varying density, in this case the atmosphere, by an infinite series of line segments each representing a single refraction. [source: *Stanford Encyclopedia of Philosophy* bio] In the whole of his argument in favor of the Earth's motion Oresme is both more explicit and much clearer than that given two centuries later by Copernicus. He was also the first to assume that color and light are of the same nature. He asserted methodological naturalism: "there is no reason to take recourse to the heavens, the last refuge of the weak, or demons, or to our glorious God as if He would produce these effects directly, more so than those effects whose causes we believe are well known to us." [source: Wikipedia: "Science in the Middle Ages"] He also showed how to interpret the difficulties encountered in "the Sacred Scriptures wherein it is stated that the sun turns, etc. It might be supposed that here Holy Writ adapts itself to the common mode of human speech, as also in several places, for instance, where it is written that God repented Himself, and was angry and calmed Himself and so on, all of which is, however, not to be taken in a strictly literal sense". Finally, Oresme offered several considerations favourable to the hypothesis of the Earth's daily motion. In order to refute one of the objections raised by the Peripatetics against this point, Oresme was led to explain how, in spite of this motion, heavy bodies seemed to fall in a vertical line; he admitted their real motion to be composed of a fall in a vertical line and a diurnal rotation identical with that which they would have if bound to the Earth. This is precisely the principle to which Galileo was afterwards to turn. He adopted Buridan's theory of dynamics in its entirety. [source: Wikipedia: "History of Physics"] "Most of the essential elements in both his [i.e., Copernicus'] criticism of Aristotle and his theory of motion can be found in earlier scholastic writers, particularly in Oresme." [source: Kuhn, p. 154]

**Henry of Langenstein** (c. 1325-1397) In 1368, at the occasion of the appearance of a comet, which the astrologers of his times claimed to be a sure foreboding of certain future events, he wrote a treatise entitled *Quæstio de cometa*, in which he refutes the then prevalent astrological superstitions. At the instance of the university he wrote three other treatises on the same subject, completed in 1373. [source: Catholic Encyclopedia bio]

**Marsilius of Inghen** (c. 1330-1396) He applied a synthesis of the new 14th century physics of Buridan, Thomas Bradwardine and Oresme in his commentaries on Aristotle. [source: Wikipedia bio]

**Nicholas of Cusa** (1401–1464; cardinal) Nicholas anticipated many later ideas in mathematics, cosmology, astronomy, and experimental science while constructing his own original version of systematic Neoplatonism. In Book II of *On Learned Ignorance* he holds that the natural universe is characterized by change or motion; it is not static in time and space. But finite change and motion, ontologically speaking, are also matters of more and less and have no fixed maximum or minimum. This "ontological relativity" leads Cusanus to some remarkable conclusions about the earth and the physical universe, based not on empirical observation but on metaphysical grounds. The earth is not fixed in place at some given point because nothing is utterly at rest; nor can it be the exact physical center of the natural universe, even if it seems nearer the center than "the fixed stars." Because the universe is in motion without fixed center or boundaries, none of the spheres of the Aristotelian and Ptolemaic world picture are exactly spherical. None of them has an exact center, and the "outermost sphere" is not a boundary. Cusanus thus shifts the typical medieval picture of the created universe toward later views, but on ontological grounds. The natural universe itself, as a contracted image of God, has a physical center that can be anywhere and a circumference that is nowhere. [source: *Stanford Encyclopedia of Philosophy* bio] Cusanus said that no perfect circle can exist in the universe (opposing the Aristotelean model, and also Copernicus' later assumption of circular orbits), thus opening the possibility for Kepler's model featuring elliptical orbits of the planets around the Sun. He made important contributions to the field of mathematics by developing the concepts of the infinitesimal and of relative motion. He was the first to use concave lenses to correct myopia. His writings were essential for Leibniz's discovery of calculus as well as Cantor's later work on infinity. [source: Wikipedia bio] The astronomical views of the cardinal are scattered through his philosophical treatises. The earth is a star like other stars [spherical], is not the centre of the universe, is not at rest, nor are its poles fixed. The celestial bodies are not strictly spherical, nor are their orbits circular. The difference between theory and appearance is explained by relative motion. [source: *Catholic Encyclopedia* bio] "Copernicus . . . had probably at least heard of the very influential treatise in which the fifteenth-century Cardinal, Nicholas of Cusa, derived the motion of the earth from the plurality of worlds in an unbounded Neoplatonic universe. The earth's motion had never been a popular concept, but by the sixteenth century it was scarcely unprecedented." [source: Kuhn, p. 144]

**Giovanni Bianchini** (1410-c. 1469) He was a professor of mathematics and astronomy at the University of Ferrara and the first mathematician in Europe to use decimal positional fractions for his trigonometric tables. In *De arithmetica*, part of the *Flores almagesti*, he uses operations with negative numbers and expresses the *Law of Signs*. [source: Wikipedia bio]

**Georg von Peuerbach** (1423-1461) He has been called the father of mathematical and observational astronomy in the West. He replaced Ptolemy's chords with the sines from Arabic mathematics, and calculated tables of sines for every minute of arc for a radius of 600,000 units. This was the first transition from the duodecimal to the decimal system. His observations were made with very simple instruments, an ordinary plumb-line being used for measuring the angles of elevation of the stars. He is credited with the invention of several scientific instruments, including the regula, the geometrical square. He authored the first printed astronomical textbook, the *Theoricae novae Planetarum* (printed in 1472). [source: Wikipedia bio]

**Regiomontanus** (1436-1476) He wrote *De Triangulis omnimodus* (*On Triangles*, 1464): one of the first textbooks presenting the current state of trigonometry, built at Nuremberg the first astronomical observatory in Germany, and worked at the Observatory of Großwardein (Oradea) in Transylvania, the first in Europe. [source: Wikipedia bio]

**Leonardo da Vinci** (1452-1519) Polymath: painter, sculptor, architect, musician, scientist, mathematician, engineer, inventor, anatomist, geologist, cartographer, botanist and writer. A keen observer, and endowed with insatiable curiosity, he had studied a great number of works, amongst which we may mention the various treatises of the School of Jordanus, various books by Albert of Saxony, and in all likelihood the works of Nicholas of Cusa; then, profiting by the learning of these scholars, he formally enunciated or else simply intimated many new ideas. The statics of the School of Jordanus led him to discover the law of the composition of concurrent forces stated as follows: the two component forces have equal moments as regards the direction of the resultant, and the resultant and one of the components have equal moments as regards the direction of the other component. The statics derived from the properties which Albert of Saxony attributed to the centre of gravity caused Vinci to recognize the law of the polygon of support and to determine the centre of gravity of a tetrahedron. He also presented the law of the equilibrium of two liquids of different density in communicating tubes, and the principle of virtual displacements seems to have occasioned his acknowledgement of the hydrostatic law known as Pascal's. Vinci continued to meditate on the properties of impetus, which he called *impeto* or *forza*, and the propositions that he formulated on the subject of this power very often showed a fairly clear discernment of the law of the conservation of energy. These propositions conducted him to remarkably correct and accurate conclusions concerning the impossibility of perpetual motion. He asserted that the velocity of a body that falls freely is proportional to the time occupied in the fall, and he understood in what way this law extends to a fall on an inclined plane. Vinci was much engrossed in the analysis of the deformations and elastic reactions which cause a body to rebound after it has struck another, and this doctrine, formulated by Buridan, Albert of Saxony, and Marsile of Inghem he applied in such a way as to draw from it the explanation of the flight of birds. This flight is an alternation of falls during which the bird compresses the air beneath it, and of rebounds due to the elastic force of this air. Until the great painter discovered this explanation, the question of the flight of birds was always looked upon as a problem in statics, and was likened to the swimming of a fish in water. Vinci attached great importance to the views developed by Albert of Saxony in regard to the Earth's equilibrium. Like the Parisian master, he held that the centre of gravity within the terrestrial mass is constantly changing under the influence of erosion and that the Earth is continually moving so as to bring this centre of gravity to the centre of the World. These small, incessant motions eventually bring to the surface of the continents those portions of earth that once occupied the bed of the ocean and, to place this assertion of Albert of Saxony beyond the range of doubt, Vinci devoted himself to the study of fossils and to extremely cautious observations which made him the creator of Stratigraphy. In many passages in his notes Vinci asserts, like Nicholas of Cusa that the moon and the other wandering stars are worlds analogous to ours, that they carry seas upon their surfaces, and are surrounded by air; and the development of this opinion led him to talk of the gravity binding to each of these stars the elements that belonged to it. On the subject of this gravity he professed a theory similar to Oresme's. Hence it would seem that, in almost every particular, Vinci was a faithful disciple of the great Parisian masters of the fourteenth century, of Buridan, Albert of Saxony, and Oresme. [source: Wikipedia: "History of Physics"] He conceptualised a helicopter, a tank, concentrated solar power, a calculator, the double hull and outlined a rudimentary theory of plate tectonics. Some of his smaller inventions, such as an automated bobbin winder and a machine for testing the tensile strength of wire, entered the world of manufacturing unheralded. As a scientist, he greatly advanced the state of knowledge in the fields of anatomy, civil engineering, optics, and hydrodynamics. He made one of the first scientific drawings of a fetus *in utero*. [source: Wikipedia bio] He wrote:

> And a little beyond the sandstone conglomerate, a tufa has been formed, where it turned towards Castel Florentino; farther on, the mud was deposited in which the shells lived, and which rose in layers according to the levels at which the turbid Arno flowed into that sea. And from time to time the bottom of the sea was raised, depositing these shells in layers, as may be seen in the cutting at Colle Gonzoli, laid open by the Arno which is wearing away the base of it; in which cutting the said layers of shells are very plainly to be seen in clay of a bluish colour, and various marine objects are found there.

This quotation makes clear the breadth of Leonardo's understanding of Geology, including the action of water in creating sedimentary rock, the tectonic action of the earth in raising the sea bed and the action of erosion in the creation of geographical features. He wrote about astronomy:

> The earth is not in the centre of the Sun's orbit nor at the centre of the universe, but in the centre of its companion elements, and united with them. And any one standing on the moon, when it and the sun are both beneath us, would see this our earth and the element of water upon it just as we see the moon, and the earth would light it as it lights us.

Leonardo was a master of mechanical principles. He

utilized leverage and cantilevering, pulleys, cranks, gears, including angle gears and rack and pinion gears; parallel linkage, momentum, centripetal force and the aerofoil. While he designed a number of man-powered flying machines with mechanical wings that flapped, he also designed a parachute and a light hang glider which could have flown. [source: Wikipedia: "Science and inventions of Leonardo da Vinci"]

**Domenico Maria Novara** (1454–1504) For 21 years he was professor of astronomy at the University of Bologna, also lectured in mathematics at Rome. He was notable as a Platonist astronomer, and in 1496 he taught Nicholas Copernicus astronomy. His teacher had been the famous astronomer Regiomontanus, who was once a pupil of Georg Peuerbach. [source: Wikipedia bio]

# Chapter Four

## 70 Catholic, Protestant and Otherwise Religious Prominent Scientists: 1500-1700 (From Copernicus to Steno, Boyle, and Ray)

Nicolaus Copernicus (1473-1543) Copernicus' epochal book, *De revolutionibus orbium coelestium* (*On the Revolutions of the Celestial Spheres*), published just before his death in 1543, is often regarded as the starting point of modern astronomy and the defining epiphany that began the scientific revolution. His heliocentric model, with the Sun at the center of the universe, demonstrated that the observed motions of celestial objects can be explained without putting Earth at rest in the center of the universe. Among the great polymaths of the Renaissance, Copernicus was a mathematician, astronomer, physician, quadrilingual polyglot, classical scholar, translator, artist, Catholic cleric, jurist, governor, military leader, diplomat and economist. In 1533, Johann Albrecht Widmannstetter delivered a series of lectures in Rome outlining Copernicus' theory. Pope Clement VII and several Catholic cardinals heard the lectures and were interested in the theory. On 1 November 1536, Cardinal Nikolaus von Schönberg, Archbishop of Capua, wrote to Copernicus from Rome, expressing great interest and approval. Copernicus' "Commentariolus" summarized his heliocentric theory. It listed the "assumptions" upon which the theory was based as follows:

1. There is no one center of all the celestial circles or spheres.

2. The center of the earth is not the center of the universe, but only of gravity and of the lunar sphere.

3. All the spheres revolve about the sun as their mid-point, and therefore the sun is the center of the universe.

4. The ratio of the earth's distance from the sun to the height of the firmament (outermost celestial sphere containing the stars) is so much smaller than the ratio of the earth's radius to its distance from the sun that the distance from the earth to the sun is imperceptible in comparison with the height of the firmament.

5. Whatever motion appears in the firmament arises not from any motion of the firmament, but from the earth's motion. The earth together with its circumjacent elements performs a complete rotation on its fixed poles in a daily motion, while the firmament and highest heaven abide unchanged.

6. What appear to us as motions of the sun arise not from its motion but from the motion of the earth and our sphere, with which we revolve about the sun like any other planet. The earth has, then, more than one motion.

7. The apparent retrograde and direct motion of the planets arises not from their motion but from the earth's. The motion of the earth alone, therefore, suffices to explain so many apparent inequalities in the heavens. [source: Wikipedia bio]

Copernicus thought that God was personally responsible for all the activity in the heavens. The regularity he was discovering in the movements of the planets was, for him, a manifestation of the faithfulness of a loving Creator. [source:

Tripp] There is evidence that he prayed the office, the Liturgy of Hours, every day of his adult life. On his deathbed his admirers brought him the astronomy books he had written, asking him to point out the most significant passages. He brushed them aside and instead asked a friend to write this epitaph:

> O Lord, I cannot ask for the faith that you gave to Paul; the mercy that you showed to Peter I dare not ask. But the grace that you showed to the dying robber, that, Lord, show to me. [source]

In his letter of dedication of his book to the pope, he stated:

> If there should chance to be any exegetes ignorant of mathematics who pretend to skill in that discipline, and dare to condemn and censure this hypothesis of mine upon the authority of some scriptural passage twisted to their purpose, I value them not, but disdain their unconsidered judgment. For it is known that [Church father] Lactantius -- a poor mathematician though in other respects a worthy author -- writes very childishly about the shape of the earth when he scoffs at those who affirm it to be a globe. Hence it should not seem strange to the ingenious if people of that sort should in turn deride me. But mathematics is written for mathematicians, by whom, if I am not deceived, these labors of mine will be recognized as contributing something to their domain, as also to that of the Church over which Your Holiness now reigns. [source: Stillman Drake, *Discoveries and Opinions of Galileo* (Doubleday Anchor Books, 1957), p. 180. This portion was later cited approvingly by Galileo, in his defense of his own heliocentric position, following Copernicus]

See also the *Stanford Encyclopedia of Philosophy* entry on Copernicus. For more on his religious views, and religion and heliocentrism, see: Joseph L. Spradley, "Tradition and Faith in the Copernican Revolution." Copernicus erred, however, in asserting circular orbits and in holding that the sun was the stationary center of the universe, with not only the earth and the other planets of the solar system, but also all the other stars, moving around it. He also believed that transparent rotating crystalline spheres carried the planets in their orbits.

**Girolamo Fracastoro** (1478-1553) In 1546 he proposed that epidemic diseases are caused by transferable tiny particles or "spores" that could transmit infection by direct or indirect contact or even without contact over long distances. In his writing, the "spores" of disease may refer to chemicals rather than to any living entities.

> I call fomites [from the Latin fomes, meaning "tinder"] such things as clothes, linen, etc., which although not themselves corrupt, can nevertheless foster the essential seeds of the contagion and thus cause infection.

His theory remained influential for nearly three centuries, before being displaced by germ theory. [source: Wikipedia bio] The British medical journal *Lancet* called Girolamo Fracastoro "the physician who did most to spread knowledge of the origin, clinical details and available treatments of [the sexually-transmitted disease syphilis] throughout a troubled Europe." His poem, *Syphilis sive morbus gallicus,* 1530, gave name to the disease. Fracastoro excelled in the arts and sciences and engaged in a lifelong study of literature, music, geography, geology, philosophy, mathematics, and astronomy, as well as medicine. [source: Holding, Scientists of the Christian Faith bio]

**Celio Calcagnini** (1479-1541) He proposed to explain the daily motion of the stars by attributing to the Earth a rotation from West to East, complete in one sidereal day. His dissertation, *Quod c lum stet, terra vero moveatur,* although seeming to have been written about 1530, was not published until 1544, when it appeared in a posthumous edition of the author's works. Calcagnini declared that the Earth, originally in equilibrium in the centre of the universe, received a first impulse which imparted to it a rotary motion, and this motion, to which nothing was opposed, was indefinitely preserved by virtue of the principle set forth by Buridan and accepted by Albert of Saxony and Nicholas of Cusa. [source: Wikipedia: "History of Physics"]

**Michael Stifel** (1486-1567) Stifel discovered logarithms and invented an early form of logarithm tables decades before John Napier. He made important innovations in mathematical notation and first used multiplication by juxtaposition (with no symbol between the terms) and the term "exponent". [source: Wikipedia bio]

**Otto Brunfels** (1488–1534) He is often called a father of botany, because, in his botanical writings, he relied not so much on the ancient authors as on his own observations, and described plants according to the latter. [source: Wikipedia bio]

**Georgius Agricola** (1494-1555) He is known as "the father of mineralogy". In 1544 he published the *De ortu et causis subterraneorum*, in which he laid the first foundations of a physical geology. In 1546 he wrote *De veteribus et novis metallis*, a comprehensive account of the discovery and occurrence of minerals and also more commonly known as *De Natura Fossilium*. His most famous work, the *De re metallica* libri xii, was published in 1556. It is considered a classic document of medieval metallurgy, unsurpassed for two centuries: a complete and systematic treatise on mining and extractive metallurgy. He describes and illustrates how ore veins occur in and on the ground, making the work an early contribution to the developing science of geology. [source: Wikipedia bio]

**Dominic Soto** (1494-1560) applied the law to the uniformly accelerated falling of lighter and heavier bodies and to the uniformly decreasing ascension of projectiles, far in advance of Galileo. [source: *Catholic Encyclopedia*: "Nicole Oresme"]

**Francesco Maurolico** (1494-1575; Benedictine abbot) He was a mathematician and astronomer who made contributions to the fields of geometry, optics, conics, mechanics, music, and astronomy. His *Arithmeticorum libri duo* (1575) includes the first known proof by mathematical induction. He published a *Cosmographia* in which he described a methodology for measuring the earth, which was later employed by Jean Picard in measuring the meridian in 1670. [source: Wikipedia bio]

**Gerolamo Cardano** (1501-1576) He was the first to describe typhoid fever. Today, he is best known for his achievements in algebra. He published the solutions to the cubic and quartic equations in his 1545 book *Ars Magna*. He acknowledged the existence of what are now called imaginary numbers, although he did not understand their properties (mathematical field theory was developed centuries later). In *Opus novum de proportionibus* he introduced the binomial coefficients and the binomial theorem. His book *Liber de ludo aleae* ("Book on Games of Chance") contains the first systematic treatment of probability. Cardano invented several mechanical devices including the combination lock, the gimbal consisting of three concentric rings allowing a supported compass or gyroscope to rotate freely, and the Cardan shaft with universal joints, which allows the transmission of rotary motion at various angles and is used in vehicles to this day. He made several contributions to hydrodynamics and held that perpetual motion is impossible, except in celestial bodies. Significantly, in the history of deaf education, he said that deaf people were capable of using their minds, argued for the importance of teaching them, and was one of the first to state that deaf people could learn to read and write without learning how to speak first. [source: Wikipedia bio]

**William Turner** (c. 1508–1568) He is sometimes called the "father of English botany" (concentrating on herbs) and was also an ornithologist. [source: Wikipedia bio]

**Realdo Colombo** (c. 1510-1559) He made several important advances in anatomy, including the discovery of the pulmonary circuit and the fact that the main action of the heart was contraction, rather than dilation. [source: Wikipedia bio]

**Bernardino Telesio** (1509-1588) He proposed an inquiry into the data given by the senses, from which he held that all true knowledge really comes. He wrote: "That the construction of the world and the magnitude of the bodies contained within it, and the nature of the world, is to be searched for not by reason as was done by the ancients, but is to be understood by means of observation." Though Francis Bacon is generally credited nowadays with the codification of an inductive method that wholeheartedly endorses observation as the primary procedure for acquiring knowledge, he was certainly not the first to suggest that sensory perception should be the primary source for knowledge. Among natural philosophers from the Renaissance, this honor is generally bestowed upon Telesio. [source: Wikipedia bio]

**Ambroise Paré** (c. 1510-1590) He is considered as one of the fathers of surgery. He was a leader in surgical techniques and battlefield medicine, especially the treatment of wounds. He was also an anatomist and the inventor of several surgical instruments. Paré also introduced the ligature of arteries instead of cauterization during amputation. To do this he designed the "Bec de Corbin" ("crow's beak"), a predecessor to modern hemostats. Although ligatures often spread infection, it still was an important breakthrough in surgical practice. Paré was also an important figure in the progress of obstetrics in the middle of the 16th century. He revived the practice of the podalic version of delivery. He contributed both to the practice of surgical amputation and to the design of limb prostheses. He also invented some ocular prostheses, making artificial eyes from enameled gold, silver, porcelain and glass. [source: Wikipedia bio]

**Andreas Vesalius** (1514-1564) He authored one of the most influential books on human anatomy, *De humani corporis fabrica* (On the Workings of the Human Body) and is often referred to as the founder of modern human anatomy. Vesalius' work on the vascular and circulatory systems was his greatest contribution to modern medicine. He defined a nerve as the mode of transmitting sensation and motion and believed that they didn't originate from the heart, but that nerves stemmed from the brain. His most significant contribution to the study of the brain was his trademark illustrations in which he depicts the corpus callosum, the thalamus, the caudate nucleus, the lenticular nucleus, the globus pallidus, the putamen, the pulvinar, and the cerebral peduncles for the first time. Due to his impressive study of the human skull and the variations of its features he is said to have been responsible for the launch of the study of physical anthropology. [source: Wikipedia bio]

**Georg Joachim Rheticus** (1514-1574) He is perhaps best known for his trigonometric tables and for being Copernicus' sole pupil. In 1551, Rheticus produced a tract titled, *Canon of the Science of Triangles*, the first publication of six-function trigonometric tables, though the term trigonometry was not used until 1595. This pamphlet was to be an introduction to Rheticus' greatest work, a full set of tables to be used in angular astronomical measurements. His student, Valentin Otto oversaw the hand computation of approximately one hundred thousand ratios to at least ten decimal places. When completed in 1596, the volume, *Opus palatinum de triangulus*, filled nearly fifteen hundred pages. Its tables of values were accurate enough to be used as the basis for astronomical computation into the early twentieth century. [source: Wikipedia bio]

**Valerius Cordus** (1515-1544) He authored one of the greatest pharmacopoeias (*Dispensatorium*) and one of the most celebrated herbals in history. He also pioneered a method for synthesizing ether which involved adding sulfuric acid to ethyl alcohol, and identified several new plant species. [source: Wikipedia bio]

**Conrad Gessner** (1516-1565) His five-volume *Historiae animalium* (1551-1558) is considered the beginning of modern zoology. To his contemporaries he was best known as a botanist, though his botanical manuscripts were not published till long after his death. [source: Wikipedia bio]

Andrea Cesalpino (1519-1603) he laid the foundation of the morphology and physiology of plants and produced the first scientific classification of flowering plants. [source: Wikipedia bio]

Gabriele Falloppio (1523-1562; canon) He added much to what was known before about the internal ear and described in detail the tympanum and its relations to the osseous ring in which it is situated. He also described minutely the circular and oval windows (fenestræ) and their communication with the vestibule and cochlea. He was the first to point out the connection between the mastoid cells and the middle ear. His description of the lacrimal ducts in the eye was a marked advance on those of his predecessors and he also gave a detailed account of the ethmoid bone and its cells in the nose. His contributions to the anatomy of the bones and muscles were very valuable. It was in myology particularly that he corrected Vesalius. He studied the reproductive organs in both sexes, and described the Fallopian tube, which leads from the ovary to the uterus and now bears his name. He was the first to use an aural speculum for the diagnosis and treatment of diseases of the ear, and his writings on surgical subjects are still of interest. [source: Wikipedia bio]

Carolus Clusius (or Charles de l'Écluse) (1526-1609) He was a pioneering botanist and perhaps the most influential of all 16th century scientific horticulturists. Clusius laid the foundations of Dutch tulip breeding and the bulb industry today. He authored *Exoticorum libri decem* (1605), an important survey of exotic flora and fauna that is still often consulted. [source: Wikipedia bio]

Julius Caesar Aranzi (1529-1589) From Aranzio came the first correct account of the anatomical peculiarities of the fetus, and he was the first to show that the muscles of the eye do not arise from the dura mater but from the margin of the optic hole. He also, after considering the anatomical relations of the cavities of the heart, the valves and the great vessels, corroborated the views of

Realdo Colombo regarding the course that the blood follows in passing from the right to the left side of the heart. Aranzio was the first anatomist to describe distinctly the inferior cornua of the ventricles of the cerebrum, who recognizes the objects by which they are distinguished, and who gives them the name by which they are still known (hippocampus) in 1564. He also was the first to discover that the blood of mother and fetus remain separate during pregnancy. [source: Wikipedia bio]

**Giambattista Benedetti** (1530-1590) His theory of motion predicted that two objects of the same material but of different weights would fall at the same speed, and also that objects of different materials in a vacuum would fall at different though finite speeds. Moreover, he held that two objects of the same material but of different surface areas would only fall at equal speeds in a vacuum: according to the then current theory of impetus. It is thought that Galileo derived his initial theory of the speed of a freely falling body from his reading of Benedetti's works. [source: Wikipedia bio]

**Christopher Clavius** (1538-1612; Jesuit priest) He was the main architect of the modern Gregorian calendar. In his last years he was probably the most respected astronomer in Europe and his textbooks were used for astronomical education for over fifty years in Europe. In logic, Clavius' Law (inferring of the truth of a proposition from the inconsistency of its negation), is named after him. [source: Wikipedia bio] The historian of science George Sarton calls him "the most influential teacher of the Renaissance". In his *Astrolabium* (Rome, 1593) he uses a dot to separate whole numbers from decimal fractions, but it would be twenty more years before the decimal point would be widely accepted. In his *Algebra* (Rome, 1608) Clavius was the first to use parenthesis to express aggregation and the first to use a symbol for an unknown quantity. Other innovations were also seen in the symbols attributed to him by Florian Cajori such as the radical sign, plus and minus signs. In his *Triangula sphaerica* (Mainz 1611) Clavius summarized all contemporary knowledge of plane and spherical trigonometry. His *prostlaphaeresis*, the grandparent of logarithms, relied on the sine of the sum and differences of numbers. [source: *Adventures of Early Jesuit Scientists* bio]

**Jose de Acosta** (1540-1600; Jesuit priest) For his work on altitude sickness in the Andes he is listed as one of the pioneers of modern aeronautical medicine. He was one of the earliest geophysicists, having been among the first to observe, record and analyze earthquakes, volcanoes, tides, currents, magnetic declinations and meteorological phenomena. He denied the commonly held opinion that earthquakes and volcanoes originated from the same cause, and offered the earliest scientific explanation of the tropical trade winds. [source: *Adventures of Early Jesuit Scientists* bio]

**William Gilbert** (1544-1603) He is remembered for his book *De Magnete* (1600), and is credited as one of the originators of the term *electricity*. He is regarded by some as the father of electrical engineering or electricity and magnetism. He determined that the Earth was itself magnetic and that this was the reason compasses point north. He was the first to argue, correctly, that the centre of the Earth was iron. Gilbert also studied static electricity using amber; amber is called elektron in Greek, so Gilbert decided to call its effect the electric force. He invented the first electrical measuring instrument, the electroscope, in the form of a pivoted needle he called the versorium. [source: Wikipedia bio]

**Tycho Brahe** (1546–1601) Tycho is credited with the most accurate astronomical observations of his time. No one before Tycho had attempted to make so many planetary observations. In 1573 he published a small book, *De nova stella*, thereby coining the term *nova* for a "new" star. After his death, his records of the motion of the planet Mars provided evidence to support Kepler's discovery of the ellipse. His system also offered a major innovation: entirely eliminating Copernicus' rotating crystalline spheres that determined orbits. Celestial objects observed near the horizon and above appear with a greater altitude than the real one, due to atmospheric refraction, and one of Tycho's most important innovations was that he worked out and published the very first tables for the systematic correction of this possible source of error. Tycho's distinctive contributions to lunar theory include his discovery of the Variation of the Moon's longitude. In addition to his contributions to astronomy, he was famous in his own time also for his contributions to medicine; his herbal medicines were in use as late as the 1900s. [source: Wikipedia bio] Tycho Brahe erred insofar as he was a geocentrist and held (Tychonic "geoheliocentric" system) that the sun and moon revolve around the earth, and the other five planets revolve around the sun: all in circular, not elliptical orbits. Also, in his system the earth did not rotate.

**Simon Stevin** (1548-1620) Stevin was the first to show how to model regular and semiregular polyhedra by delineating their frames in a plane. He also distinguished stable from unstable equilibria. Stevin proved the law of the equilibrium on an inclined plane, using an ingenious and intuitive diagram showing a rope containing evenly spaced beads draped over an inclined plane (see the illustration on the side). Physicist Richard Feynman mentions with reverence in his Lectures on Physics that with the diagram, Stevin elegantly proves the law of conservation of energy. Stevin discovered the hydrostatic paradox, which states that the downward pressure of a liquid is independent of the shape of the vessel, and depends only on its height and base. In 1586, he demonstrated that two objects of different weight fall down with exactly the same acceleration. Decimal fractions had been employed for the extraction of square roots centuries before his time by Islamic mathematicians but nobody established their daily use before Stevin. His general notion of a real number was later broadly accepted. [source: Wikipedia bio]

**John Napier** (1550-1617) He is the inventor of the so-called "Napier's bones". Napier also made common the use of the decimal point in arithmetic and mathematics. He advanced theorems in Spherical Trigonometry, usually known as Napier's Rules of Circular Parts. [source: Wikipedia bio]

**Bartholomaeus Pitiscus** (1561-1613) He first coined the word *trigonometry* and is sometimes credited with inventing the decimal point. [source: Wikipedia bio]

**Francis Bacon** (1561-1626) His works established and popularized deductive methodologies for scientific inquiry, often called the *Baconian method* or simply, the scientific method, consisting of procedures for isolating the form, nature or cause of a phenomenon, employing the method of agreement, method of difference, and method of concomitant variation devised by Avicenna in 1025. His demand for a planned procedure of investigating all things natural marked a new turn in the rhetorical and theoretical framework for science, much of which still surrounds conceptions of proper methodology today. [source: Wikipedia bio] Bacon suggests that you draw up a list of all things in which the phenomenon you are trying to explain occurs, as well as a list of things in which it does not occur. Then you rank your lists according to the degree in which the phenomenon occurs in each one. Then you should be able to deduce what factors match the occurrence of the phenomenon in one list and don't occur in the other list, and also what factors change in accordance with the way the data had been ranked. From this Bacon concludes you should be able to deduce by elimination and inductive reasoning what is the cause underlying the phenomenon. [source: Wikipedia: "Baconian method"] He repudiates the syllogistic method and defines his alternative procedure as one "which by slow and faithful toil gathers information from things and brings it into understanding". Induction implies ascending to axioms, as well as a descending to works, so that from axioms new particulars are gained and from these new axioms. The inductive method starts from sensible experience and moves via natural history (providing sense-data as guarantees) to lower axioms or propositions, which are derived from the tables of presentation or from the abstraction of notions. From the more general axioms Bacon strives to reach more fundamental laws of nature (knowledge of forms), which lead to practical deductions as new experiments or works. Bacon came to the fundamental insight that *facts* cannot be collected from nature, but must be constituted by methodical procedures, which have to be put into practice by scientists in order to ascertain the empirical basis for inductive generalizations. His induction, founded on collection, comparison, and exclusion of factual qualities in things and their interior structure, proved to be a revolutionary achievement within natural philosophy, for which no example in classical antiquity existed. Above all, his emphasis on negative instances for the procedure of induction itself can claim a high importance with regard to knowledge acquisition and has been acclaimed as an innovation by scholars of our time. Some have detected in Bacon a forerunner of Karl Popper in respect of the method of falsification. Finally, it cannot be denied that Bacon's methodological program of induction includes aspects of deduction and abstraction on the basis of negation and exclusion. Contemporary scholars have praised his inauguration of the theory of induction. [source: *Stanford Encyclopedia of Philosophy* bio] Bacon stated:

> It is true, that a little philosophy inclineth man's mind to atheism; but depth in philosophy bringeth men's minds about to religion. For while the mind of man looketh upon second causes scattered, it may sometimes rest in them, and go no further; but when it beholdeth the chain of them, confederate and linked together, it must needs fly to Providence and Deity.

**Galileo Galilei** (1564-1642) Galileo has been called the "father" of modern observational astronomy, of modern physics, of science, and of modern science. Stephen Hawking says, "Galileo, perhaps more than any other single person, was responsible for the birth of modern science." Einstein was of the same opinion. The motion of uniformly accelerated objects, taught in nearly all high school and introductory college physics courses, was studied by Galileo as the subject of kinematics. He derived the correct kinematical law for the distance travelled during a uniform acceleration starting from rest—namely, that it is proportional to the square of the elapsed time. His observations about inertia were the first time that it had been mathematically expressed, verified experimentally, with the corresponding idea of frictional force, the key breakthrough in validating inertia. Galileo is perhaps the first to clearly state that the laws of nature are mathematical. His contributions to observational astronomy include the observation and analysis of sunspots. Galileo also worked in applied science and technology, inventing an improved military compass and other instruments. Galileo constructed a thermometer, using the expansion and contraction of air in a bulb to move water in an attached tube. By 1624 he had perfected a compound microscope. In his last year, when totally blind, he designed an escapement mechanism for a pendulum clock. Galileo is credited with being one of the first to understand sound frequency. By scraping a chisel at different speeds, he linked the pitch of the sound produced to the spacing of the chisel's skips, a measure of frequency. He put forward the basic principle of relativity, that the laws of physics are the same in any system that is moving at a constant speed in a straight line, regardless of its particular speed or direction. Hence, there is no absolute motion or absolute rest. This principle provided the basic framework for Newton's laws of motion and is central to Einstein's special theory of relativity. Based only on uncertain descriptions of the first practical telescope, invented by Hans Lippershey in the Netherlands in 1608, Galileo, in the following year, made a telescope with about 3x magnification. He later made improved versions with up to about 30x magnification. His observations and discovery of the four largest satellites of Jupiter created a revolution in astronomy that reverberates to this day: a planet with smaller planets orbiting it did not conform to the principles of Aristotelian Cosmology, which held that all heavenly bodies should circle the Earth. Galileo observed that Venus exhibited a full set of phases similar to that of the Moon. This discovery was arguably his most empirically practically influential contribution to the two-stage transition from full geocentrism to full heliocentrism via geo-heliocentrism. Galileo observed the Milky Way, previously believed to be nebulous, and found it to be a multitude of stars packed so densely that they appeared to be clouds from Earth. Galileo argued that stars were suns, and that they were not arranged in a spherical shell surrounding the solar system but rather were at varying distances from Earth. Brighter stars were closer suns, and fainter stars were more distant suns. His early works in dynamics, the science of motion and mechanics were his 1590 *Pisan De Motu* (*On Motion*) and his circa 1600 *Paduan Le Meccaniche* (*Mechanics*). The former was based on Aristotelian-Archimedean fluid dynamics and held that the speed of gravitational fall in a fluid medium was proportional to the excess of a body's specific weight over that of the medium, whereby in a vacuum bodies would fall with speeds in proportion to their specific weights. It also subscribed to the Hipparchan-Philoponan impetus dynamics in which impetus is self-dissipating and free-fall in a vacuum would have an essential terminal speed according to specific weight after an initial period of acceleration. [source: Wikipedia bio] Galileo accepted the inerrancy of Scripture; but he was also mindful of Cardinal Baronius's quip that the Bible "is intended to teach us how to go to heaven, not how the heavens go." And he pointed out correctly that both St. Augustine and St. Thomas Aquinas taught that the sacred writers in no way meant to teach a system of astronomy. St. Augustine wrote that:

> One does not read in the Gospel that the Lord said: I will send you the Paraclete who will teach you about the course of the sun and moon. For He willed to make them Christians, not mathematicians.

Unfortunately, there are still today biblical fundamentalists, both Protestant and Catholic, who do not understand this simple point: the Bible is not a scientific treatise. [source: George Sim Johnston, "The Galileo Affair"] In 1615 Galileo wrote a letter outlining his views to Madame Christina of Lorraine, the Grand Duchess of Tuscany, "Concerning the Use of Biblical Quotations in Matters of Science." He writes, "I think in the first place that it is very pious to say and prudent to affirm that the Holy Bible can never speak untruth—whenever its true meaning is understood." He cited Copernicus in the same vein: "He [Copernicus] did not ignore the Bible, but he knew very well that if his doctrine were proved, then it could not contradict the

Scripture when they were rightly understood". He quotes Augustine relating true reason to Scriptural truth:

> And in St. Augustine [in the seventh letter to Marcellinus] we read: 'If anyone shall set the authority of Holy Writ against clear and manifest reason, he who does this knows not what he has undertaken; for he opposes to the truth not the meaning of the Bible, which is beyond his comprehension, but rather his own interpretation; not what is in the Bible, but what he has found in himself and imagines to be there'. . . .
> 
> Copernicus [in his controversial book] never discusses matters of religion or faith, nor does he use arguments that depend in any way upon the authority of sacred writings which he might have interpreted erroneously. He stands always upon physical conclusions pertaining to the celestial motions, and deals with them by astronomical and geometrical demonstrations, founded primarily on sense experiences and very exact observations. He did not ignore the Bible, but he knew very well that if his doctrine were proved, then it could not contradict the Scriptures when they were rightly understood. [cites Copernicus: the portion is quoted in the Copernicus section above] . . . various authorities from the Bible, from theologians, and from Church Councils . . . Since I hold these to be of supreme authority, I consider it rank temerity for anyone to contradict them -- when employed according to the usage of the holy Church. . . . I neither intend nor pretend to gain from it [his book] any fruit that is not pious and Catholic.

[source: Stillman Drake, *Discoveries and Opinions of Galileo* (Doubleday Anchor Books, 1957), pp. 179-181, 186; for much more along these lines, see the full text of this letter -- derived from the same book -- , which contains superbly informed principles of (widely and frequently misunderstood) biblical hermeneutics regarding matters that have some relation to scientific observation]

Galileo again wrote in 1615:

> . . . since no two truths can contradict one another, this [Copernicus' position] and the Bible must be perfectly harmonious. [source: Drake, *ibid.*, p. 166]

> [for related reading, see: "Galileo and Biblical Exegesis" [PDF], by William E. Carroll [also in html] and "Galileo and the Bible," by Paul Newall, and "Galileo's Religion," Olaf Pedersen (1985) ]

For treatments of the "Galileo affair" and conflict with the Church, from a heavily documented Catholic perspective (there are two sides to every story, after all), see:

> *Catholic Encyclopedia*, "Galileo Galilei"
> The Galileo Controversy (Catholic Answers)
> Why Did the Catholic Church Condemn Galileo? (Kenneth J. Howell, *This Rock*, May-June 2003)
> Galileo and the Catholic Church (Robert P. Lockwood)
> Galileo (Anne W. Carroll)
> The Galileo Legend (Thomas Lessl)
> Galileo and the Magisterium: a Second Look (Jeffrey A. Mirus)
> Twisting the Knife (Wil Milan, *This Rock*, Nov-Dec 1999)
> Galileo Galilei (Bertrand Conway)
> 1633 Letter Resolves the Legend About the Galileo Case, Says Vatican Aide: Urban VIII Was Sensitive Toward Astronomer's Health, Document Indicates (*Zenit*, 21 August 2003)
> The Legacy of Galileo Galilei: Conference Discusses Scientist's Continuing Influence (Edward Pentin, *Zenit*, 3 December 2009)

Vatican to Publish New Volume on Galileo: To Include 20 Documents Discovered Since '91 (*Zenit*, 2 June 2009)
Pope John Paul II's Address Regarding Galileo (*L'Osservatore Romano*, 4 November 1992)
From Warpath to Wholeness: The Condemnation and Rehabilitation of Galileo Galilei (Mathew Chandrankunnel)
Actual documents of Galileo's trials (Vatican Archives)
Maurice A. Finocchiaro, "The Church and Galileo," *The Catholic Historical Review* (Vol. 94, No. 2, April 2008, pp. 260-282)
Galileo Revisited (Fr. Paschal)
Seven lengthy treatises on the trial of Galileo [all PDF; not necessarily all written by Catholics]
The Galileo Incident (Russell Maatman)

**Johannes Kepler** (1571-1630) the astronomer Kepler was a devout (Lutheran) Christian. His discovery of the three laws of planetary motion (including elliptical orbits) laid the foundation for Newton's theory of gravity. He regarded his study of the physical universe as "thinking God's thoughts after him" and that "the tongue of God and the finger of God cannot clash." In *The Secret of the Universe* he wrote:

> Here we are concerned with the book of nature, so greatly celebrated in sacred writings. It is in this that Paul proposes to the Gentiles that they should contemplate God like the Sun in water or in a mirror. Why then as Christians should we take any less delight in its contemplation, since it is for us with true worship to honor God, to venerate him, to wonder at him? The more rightly we understand the nature and scope of what our God has founded, the more devoted the spirit in which that is done.

Kepler felt himself to be "a high priest of the book of nature, religiously bound to alter not one jot or tittle of what it had pleased God to write down in it." And he stated:

Since we astronomers are priests of the highest God in regard to the book of nature, it befits us to be thoughtful, not of the glory of our minds, but rather, above else, of the glory of God. [source: Tripp]

Finding that an elliptical orbit fit the Mars data, he immediately concluded that all planets move in ellipses, with the sun at one focus—Kepler's first law of planetary motion. The second law is planets sweep out equal areas in equal times. The third law is "The square of the orbital period of a planet is directly proportional to the cube of the semi-major axis of its orbit." Kepler also proposed a force-based theory of lunar motion: "In Terra inest virtus, quae Lunam ciet" ("There is a force in the earth which causes the moon to move") and a new method for measuring lunar eclipses. His *Astronomiae Pars Optica* (*The Optical Part of Astronomy*) described the inverse-square law governing the intensity of light, reflection by flat and curved mirrors, and principles of pinhole cameras, as well as the astronomical implications of optics such as parallax and the apparent sizes of heavenly bodies. He also extended his study of optics to the human eye, and is generally considered by neuroscientists to be the first to recognize that images are projected inverted and reversed by the eye's lens onto the retina. This book is generally recognized as the foundation of modern optics. Kepler set out the theoretical basis of double-convex converging lenses and double-concave diverging lenses—and how they are combined to produce a Galilean telescope—as well as the concepts of real vs. virtual images, upright vs. inverted images, and the effects of focal length on magnification and reduction. He also described an improved refracting telescope—now known as the astronomical or Keplerian telescope—in which two convex lenses can produce higher magnification than Galileo's combination of convex and concave lenses. He published the first description of the hexagonal symmetry of snowflakes. In the 1930s and 1940s Koyré, and a number of others in the first generation of professional historians of science, described the "Scientific Revolution" as the central event in the history of science, and Kepler as a (perhaps the) central figure in the revolution. Koyré placed Kepler's theorization, rather than his empirical work, at the center of the intellectual transformation from ancient to modern world-views. Influential philosophers of science—such as Charles Sanders Peirce, Norwood Russell Hanson, Stephen Toulmin, and Karl Popper—have repeatedly turned to Kepler: examples of incommensurability, analogical reasoning, falsification, and many other philosophical concepts have been found in Kepler's work [source: Wikipedia bio] Kepler was correct in asserting elliptical orbits of the planets around the sun, at varying speeds (both notions having been foreseen by Nicholas of Cusa in the 15th century), but continued to err in thinking that the sun was the center of the entire universe.

**Christopher Scheiner** (1575-1650; Jesuit priest) He discovered sunspots independently of Galileo and explained the elliptical form of the sun near the horizon as the effect of refraction. In his *Oculus* (1619) he showed that the retina is the seat of vision. In 1613 Scheiner in turn contributed to the perfection of the refracting telescope with which we are familiar today. [source: *Adventures of Early Jesuit Scientists* bio]

**Benedetto Castelli** (1577-1644; Benedictine abbot) He was specially interested in the mathematical sciences and their application to hydraulics. Galileo, his teacher, and Toricelli, one of his pupils, speak very highly of his scientific attainments, and both of them frequently asked his advice. He wrote an important work on the "Mensuration of Running Water", in which he observed that the speed of a current varies inversely as the area of its cross section, and that the discharge from a vessel depends on the depth of the tap below the free surface of the water. [source: *Catholic Encyclopedia* bio]

**William Harvey** (1578-1657) The first to describe correctly and in detail the systemic circulation and properties of blood being pumped to the body and lungs by the heart, based on scientific observation. He also made a detailed analysis of the overall structure of the heart and the arteries, showing how their pulsation depends upon the contraction of the left ventricle, while the contraction of the right ventricle propels its charge of blood into the pulmonary artery. Whilst doing this, the physician reiterates the fact that these two ventricles move together almost simultaneously. He discerned the existence of the Ductus Arteriosus and explained its relative function. He reckoned the actual quantity of blood passing through the heart from the veins to the arteries and estimated the capacity of the heart, how much blood is expelled through each pump of the heart, and the amount of times the heart beats in a half an hour. He proved how the blood circulated in a circle by means of countless experiments initially done on serpents and fish. [source: Wikipedia bio]

**Jan Baptist van Helmont** (1579-1644) Sometimes considered to be "the founder of pneumatic chemistry" and introduced the word *gas* (from the Greek word *chaos*) into the vocabulary of scientists. He was the first to understand that there are gases distinct in kind from atmospheric air. He concluded that digestion was aided by a chemical reagent, or "ferment" within the body, such as inside the stomach. Harré suggests that in this way, van Helmont's idea was "very near to our modern concept of an enzyme." [source: Wikipedia bio]

**Nicolas Zucchi** (1586-1670; Jesuit priest) He constructed an apparatus in which an ocular lens was used to observe the image produced by reflection from a concave metal mirror. This was one of the earliest reflecting telescopes, in which the enlargement is obtained by the interaction of mirrors and lenses. Later, in *Optica philosophia* . . ., Zucchi described the apparatus, from which. wittingly or not, the most improved models of a slightly later date were derived. [source: Adventures of Early Jesuit Scientists bio]

**Johann Baptist Cysat** (c. 1587-1657; Jesuit priest) Cysat determined that comets circled around the sun, in parabolic, not circular orbits. He was the first to describe cometary nuclei, and was able to track the progression of the nucleus from a solid shape to one filled with starry particles. [source: Wikipedia bio]

**Giovanni Battista Zupi** (c. 1590-1650; Jesuit priest) The first person to discover that the planet Mercury had orbital phases, just like the Moon and Venus. His observations demonstrated that the planet orbited around the Sun. [source: Wikipedia bio]

**Pierre Gassendi** (1592-1655; priest) Active observational scientist, who published the first data on the transit of Mercury in 1631. Richard Popkin indicates that Gassendi was one of the first thinkers to formulate the modern "scientific outlook", of moderated scepticism and empiricism. In 1621, he was the first person to give the *Aurora Borealis* a name. [source: Wikipedia bio]

**René Descartes** (1596-1650) Often dubbed the "Father of Modern Philosophy". He is credited also as the father of analytical geometry and was one of the key figures in the Scientific Revolution. Descartes is often regarded as the first thinker to provide a philosophical framework for the natural sciences as these began to develop. As the inventor of the Cartesian coordinate system, Descartes founded analytic geometry, the bridge between algebra and geometry, crucial to the discovery of infinitesimal calculus and analysis. Descartes' theory provided the basis for the calculus of Newton and Leibniz, by applying infinitesimal calculus to the tangent line problem, thus permitting the evolution of that branch of modern mathematics. Descartes created analytic geometry, and discovered an early form of the law of conservation of momentum (the term momentum refers to the momentum of a force). Descartes also made contributions to the field of optics and discovered the law of reflection. One of Descartes' most enduring legacies was his development of Cartesian geometry which uses algebra to describe geometry. He also "invented" the notation which uses superscripts to show the powers or exponents, for example the 4 used in $x^4$ to indicate squaring of squaring. He claimed to be a devout Catholic, explaining that one of the purposes of the *Meditations* was to defend the Christian faith. Stephen Gaukroger's biography of Descartes reports that "he had a deep religious faith as a Catholic, which he retained to his dying day." [source: Wikipedia bio] Descartes wrote in his Meditations (V):

> And thus I very clearly see that the certitude and truth of all science depends on the knowledge alone of the true God, insomuch that, before I knew him, I could have no perfect knowledge of any other thing. And now that I know him, I possess the means of acquiring a perfect knowledge respecting innumerable matters, as well relative to God himself and other intellectual objects as to corporeal nature.

**Giovanni Battista Riccioli** (1598-1671; Jesuit priest) The first person to measure the rate of acceleration of a freely falling body. [source: Wikipedia bio]

**Gilles Personne de Roberval** (1602-1675) Mathematician who worked on the quadrature of surfaces and the cubature of solids, which he accomplished, in some of the simpler cases, by an original method which he called the "Method of Indivisibles". Another of Roberval's discoveries was a very general method of drawing tangents, by considering a curve as described by a moving point whose motion is the resultant of several simpler motions. He also discovered a method of deriving one curve from another, by means of which finite areas can be obtained equal to the areas between certain curves and their asymptotes. He also wrote a work on the system of the universe, in which he supports the Copernican heliocentric system and attributes a mutual attraction to all particles of matter and also the invention of a special kind of balance, the *Roberval Balance*. [source: Wikipedia bio]

**Jacques de Billy** (1602-1679; Jesuit priest) Billy produced a number of results in number theory which have been named after him. He was one of the first scientists to reject the role of astrology in science. He also rejected old notions about the malevolent influence of comets. [source: Wikipedia bio]

**Athanasius Kircher** (1602–1680; Jesuit priest) an early study of Egyptian hieroglyphs, correctly establishing the link between the ancient Egyptian language and the Coptic language, for which he has been considered the founder of Egyptology. Kircher's work with geology included studies of volcanos and fossils. He was one of the first people to observe microbes through a microscope, and was ahead of his time in proposing that the plague was caused by an infectious microorganism and in suggesting effective measures to prevent the spread of the disease. Kircher invented a magnetic clock, various automatons and the first megaphone. *The Encyclopædia Britannica* calls him a "one-man intellectual clearing house". One modern scholar, Alan Cutler, described Kircher as "one of the last thinkers who could rightfully claim all knowledge as his domain". [source: Wikipedia bio]

**Otto von Guericke** (1602-1686) His major scientific achievement was the establishment of the physics of vacuums. In 1650 he invented a vacuum pump consisting of a piston and an air gun cylinder with two-way flaps designed to pull air out of whatever vessel it was connected to, and used it to investigate the properties of the vacuum in many experiments. Guericke demonstrated the force of air pressure with dramatic experiments. He had joined two copper hemispheres of 51 cm diameter (Magdeburg hemispheres) and pumped the air out of the enclosure. Then he harnessed a team of eight horses to each hemisphere and showed that they were not able to separate the hemispheres. When air was again let into the enclosure, they were easily separated. With his experiments Guericke disproved the long-held hypothesis of "horror vacui", that nature abhors a vacuum. He proved that substances were not pulled by a vacuum, but were pushed by the pressure of the surrounding fluids. He applied the barometer to weather prediction and thus prepared the way for meteorology and invented the first electrostatic generator. [source: Wikipedia bio]

**Juan Caramuel y Lobkowitz** (1606-1682; Cistercian and archbishop) He wrote 262 works, including treatises on mathematics, astronomy, architecture, physics, logic, metaphysics, and theology. His mathematical work centered on combinatorics and he was one of the early writers on probability. [source: Wikipedia bio]

**Pierre de Fermat** (c. 1607-1665) He discovered an original method of finding the greatest and the smallest ordinates of curved lines, which is analogous to that of the then unknown differential calculus, as well as his research into the theory of numbers. He made notable contributions to analytic geometry, probability, and optics. He is best known for Fermat's Last Theorem. Fermat was the first person known to have evaluated the integral of general power functions. Using an ingenious trick, he was able to reduce this evaluation to the sum of geometric series. The resulting formula was helpful to Newton, and then Leibniz, when they independently developed the fundamental theorem of calculus. He discovered the little theorem, invented Fermat's factorization method - as well as the proof technique of infinite descent. Fermat and Pascal helped lay the fundamental groundwork for the theory of probability. From this brief but productive collaboration on the problem of points, they are now regarded as joint founders of probability theory. He a key figure in the historical development of the fundamental principle of least action in physics. Together with René Descartes, Fermat was one of the two leading mathematicians of the first half of the 17th century. Independently of Descartes, he discovered the fundamental principles of analytic geometry. [source: Wikipedia bio]

**Evangelista Torricelli** (1608-1647) Invented the barometer. Torricelli also discovered Torricelli's Law, regarding the speed of a fluid flowing out of an opening, which was later shown to be a particular case of Bernoulli's principle. Torricelli gave the first scientific description of the cause of wind: "winds are produced by differences of air temperature, and hence density, between two regions of the earth." [source: Wikipedia bio]

Giovanni Alfonso Borelli (1608-1679) His major scientific achievements are focused around his investigation into biomechanics. He related animals to machines and utilized mathematics to prove his theories. He first suggested that "muscles do not exercise vital movement otherwise than by contracting." He was also the first to deny corpuscular influence on the movements of muscles. This was proven through his scientific experiments demonstrating that living muscle did not release corpuscles into water when cut. He likened the action of the heart to that of a piston. For this to work properly he derived the idea that the arteries have be elastic. [source: Wikipedia bio]

André Tacquet (1612-1660; Jesuit priest) His work prepared ground for the eventual discovery of the calculus. He helped articulate some of the preliminary concepts necessary for Isaac Newton and Gottfried Leibniz to recognize the inverse nature of the quadrature and the tangent. He was one of the precursors of the infinitesimal calculus, developed by John Wallis. [source: Wikipedia bio]

Franciscus Sylvius (1614-1672) Founded the first academic chemical laboratory and also the Iatrochemical School of Medicine, according to which all life and disease processes are based on chemical actions. That school of thought attempted to understand medicine in terms of universal rules of physics and chemistry. Sylvius also introduced the concept of chemical affinity as a way to understand the way the human body uses salts and contributed greatly to the understanding of digestion and of bodily fluids. [source: Wikipedia bio]

John Wallis (1616-1703) Made significant contributions to trigonometry, calculus, geometry, and the analysis of infinite series, introduced the term "continued fraction" and laid down the principle of interpolation. [source: Wikipedia bio]

Francesco Maria Grimaldi (1618-1663; Jesuit priest) He was the first to make accurate observations on the diffraction of light and coined the word "diffraction". Later physicists used his work as evidence that light was a wave, and Isaac Newton used it to arrive at his more comprehensive theory of light. [source: Wikipedia bio]

**Jean-Felix Picard** (1620-1682; Jesuit priest) He was the first person to measure the size of the Earth to a reasonable degree of accuracy. He achieved this by measuring one degree of latitude along the Paris Meridian using triangulation along thirteen triangles stretching from Paris to the clocktower of Sourdon, near Amiens. His measurements produced a result of 110.46 km for one degree of latitude, which gives a corresponding terrestrial radius of 6328.9 km. The polar radius has now been measured at just over 6357 km (only a 0.44% difference). He discovered mercurial phosphorescence upon his observance of the faint glowing of a barometer. This discovery led to Newton's studies of spectrometry. Picard also developed what became the standard method for measuring the right ascension of a celestial object. In this method, the observer records the time at which the object crosses the observer's meridian. [source: Wikipedia bio]

**Edme Mariotte** (c. 1620-1684; priest) He was one of the first members of the French Academy of Sciences founded at Paris in 1666. The first volume of the *Histoire et mémoires de l'Académie* (1733) contains many original papers by him upon a great variety of physical subjects, such as the motion of fluids, the nature of color, the notes of the trumpet, the barometer, the fall of bodies, the recoil of guns, the freezing of water etc. He also did systematic study on rainbows, halos, parhelia, diffraction, and the more purely physiological phenomena of color. In 1660 he discovered the eye's blind spot. He was one of the fathers of hydrology. [source: Wikipedia bio]

**Blaise Pascal** (1623-1662) He made important contributions to the study of fluids, and clarified the concepts of pressure and vacuum. He invented the mechanical calculator (Pascal's calculator) and helped create two major new areas of research: projective geometry and probability theory. His work on the calculus of probabilities laid important groundwork for Leibniz' formulation of the infinitesimal calculus. Pascal's work in the fields of the study of hydrodynamics and hydrostatics centered on the principles of hydraulic fluids. His inventions include the hydraulic press (using hydraulic pressure to multiply force) and the syringe. Later in his life he wrote the famous Pensées: a defense of the Christian faith. [source: Wikipedia bio]

**Giovanni Domenico Cassini** (1625-1712) He was the first to observe four of Saturn's moons. In addition he discovered the Cassini Division in the rings of Saturn (1675). He shares with Robert Hooke credit for the discovery of the Great Red Spot on Jupiter and was also the first to observe differential rotation within Jupiter's atmosphere. Cassini measured for the first time the true dimensions of the solar system and was the first to make successful measurements of longitude by the method suggested by Galileo, using eclipses of the satellites of Jupiter as a clock. He made the first topographic map of an entire country (France). [source: Wikipedia bio]

**Robert Boyle** (1627-1691) The "father of chemistry". He did important work in the discovery of the part taken by air in the propagation of sound, investigations on the expansive force of freezing water, on specific gravities and refractive powers, on crystals, on electricity, on color, and on hydrostatics. [source: Wikipedia bio] Boyle was one of the leading intellectual figures of the seventeenth century. He viewed his theological interests and his work in natural philosophy as forming a seamless whole and constantly used results from the one area to enlighten matters in the other. [source: *Stanford Encyclopedia of Philosophy* bio] He left the sum of £50 per annum in his will for a series of eight lectures to be given against unbelievers. [source: Tripp] He wrote in his last will and testament: "Wish [the Royal Society, a group of scientists] a happy success in their laudable Attempts, to discover the Nature of the Works of God, and prayer that they and all other Searchers into Physical Truths, may Cordially refer their Attainments to the Glory of the Great Author of Nature, and to the Comfort of Mankind." Boyle's law states that when the temperature of a gas is constant, the volume of an enclosed gas varies inversely with pressure. [source: Snow] For more on Boyle's Christian faith, see: David L. Woodall, "The Relationship between Science and Scripture in the Thought of Robert Boyle."

**John Ray** (1627-1705) "the father of English natural history." He published important works on botany, zoology, and natural theology. His classification of plants according to similarities and differences and observation, in his *Historia Plantarum*: an important step towards modern taxonomy. He was the first to give a biological definition of the term *species*. [source: Wikipedia bio] He wrote:

> The treasures of nature are inexhaustible. . . . If man ought to reflect upon his Creator the glory of all his works, then ought he to take notice of them all and not to think anything unworthy of his cognisance. [source: Tripp]

For more on the relationship of his theology and scientific views, see: John R. Armstrong, "Rediscovering John Ray."

**Marcello Malpighi** (1628-1694) In observing the anatomy of the lung of a frog he discovered capillaries and was the first to recognize the link between arteries and veins. Malpighi used the microscope for studies on skin, kidney, and for the first interspecies comparison of the liver. He greatly extended the science of embryology. The use of microscopes enabled him to describe the development of the chick in its egg, and discovered that insects (particularly, the silk worm) do not use lungs to breathe, but small holes in their skin called tracheae. Later he falsely concluded that plants had similar tubules. He is regarded as the founder of microscopic anatomy and the first histologist. He may have been the first person to see red blood cells under a microscope. He also studied the anatomy of a brain and concluded that this organ is a gland. In terms of modern endocrinology this deduction is correct because neurotransmitter substances represent paracrine hormones, and the hypothalamus of the brain has long been recognized for its hormone-secreting capacity. He was also among the first to study human fingerprints. His book *Anatomia Plantarum* (1671) was the most exhaustive study of botany at the time. [source: Wikipedia bio]

**Christiaan Huygens** (1629-1695) His work included early telescopic studies elucidating the nature of the rings of Saturn and the discovery of its moon Titan, the invention of the pendulum clock and other investigations in timekeeping, and studies of both optics and the centrifugal force. Huygens achieved note for his argument that light consists of waves, now known as the Huygens–Fresnel principle, which became instrumental in the understanding of wave-particle duality. The interference experiments of Thomas Young vindicated Huygens' wave theory in 1801. He generally receives credit for his discovery of the centrifugal force, the laws for collision of bodies, for his role in the development of modern calculus and his original observations on sound perception (see repetition pitch). Huygens is seen as the first theoretical physicist as he was the first to use formulae in physics. He wrote the first book on probability theory, *De ratiociniis in ludo aleae* (*On Reasoning in Games of Chance*) and formulated what is now known as the second law of motion of Isaac Newton in a quadratic form. Newton reformulated and generalized that law. In 1659 Huygens derived the now well-known formula for the centripetal force. Huygens carried out experiments with internal combustion. He designed a basic form of internal combustion engine, fueled by gunpowder. [source: Wikipedia bio]

**Isaac Barrow** (1630-1677) Played a key early role in the development of infinitesimal calculus; in particular, the discovery of the fundamental theorem of calculus. His work centered on the properties of the tangent; Barrow was the first to calculate the tangents of the kappa curve. He was also renowned for his sermons, and he wrote treatises such as his *Expositions of the Creed*, *The Lord's Prayer*, *Decalogue*, and *Sacraments*. [source: Wikipedia bio]

**Jean Richer** (1630-1696). He first determined the distance of Earth from the sun and the shape of the Earth (a spheroid flattened at the poles). His measurements of the orbit of Mars contributed to the first accurate calculations of the size and orbits of the planets of the solar system. [source: Holding, *Scientists of the Christian Faith*] He also measured the length of the seconds pendulum, that is a pendulum with a swing of one second, and found that in Cayenne it was 1.25 *lignes* (2.8 millimeters) shorter than at Paris. This was due to the increase of gravitational force with latitude, due to the oblate shape of the Earth. He thus became the first person to observe a change in gravitational force, beginning the science of gravimetry. [source: Wikipedia bio]

**Francesco Lana de Terzi** (c. 1631-1687; Jesuit priest) He has been referred to as the Father of Aeronautics for his pioneering efforts, turning the aeronautics field into a science by establishing "a theory of aerial navigation verified by mathematical accuracy". He also developed the idea that developed into Braille. [source: Wikipedia bio]

**Robert Hooke** (1635-1703) He is known for his law of elasticity (Hooke's law), his book, *Micrographia*, and for first applying the word "cell" to describe the basic unit of life. He built some of the earliest Gregorian telescopes, observed the rotations of Mars and Jupiter, and, based on his observations of fossils, was an early proponent of biological evolution. He investigated the phenomenon of refraction, deducing the wave theory of light, and was the first to suggest that matter expands when heated and that air is made of small particles separated by relatively large distances. He also came near to deducing that gravity follows an inverse square law, and that such a relation governs the motions of the planets, an idea which was subsequently developed by Newton. Hooke's work on elasticity culminated, for practical purposes, in his development of the balance spring or hairspring, which for the first time enabled a portable timepiece – a watch – to keep time with reasonable accuracy. Hooke argued for an attracting principle of gravitation in *Micrographia* (1665). Hooke's 1666 Royal society lecture "On gravity" added two further principles – that all bodies move in straight lines till deflected by some force and that the attractive force is stronger for closer bodies. He also was an early observer of the rings of Saturn. [source: Wikipedia bio]

**James Gregory** (1638-1675) In his *Optica Promota*, he describes the first practical reflecting telescope: now called the Gregorian telescope. His novel idea was to use both mirrors and lenses in his telescope. He showed that the combination would work more effectively than a telescope which used only mirrors or used only lenses. Gregory anticipated Newton in discovering both the interpolation formula and the general binomial theorem as early as 1670; he discovered Taylor expansions more than 40 years before Taylor; he solved Kepler's famous problem of how to divide a semicircle by a straight line through a given point of the diameter in a given ratio (his method was to apply Taylor series to the general cycloid); he gives one of the earliest examples of a comparison test for convergence, essentially giving Cauchy's ratio test, together with an understanding of the remainder; he gave a definition of the integral which is essentially as general as that given by Riemann; his understanding of all solutions to a differential equation, including singular solutions, is impressive; he appears to be the first to attempt to prove that π and e are not the solution of algebraic equations; he knew how to express the sum of the nth powers of the roots of an algebraic equation in terms of the coefficients. [source: Mactutor bio]

**Blessed Nicolas Steno** (1638-1686; bishop) He is considered the father of geology. Steno, in his *Dissertationis prodromus* of 1669 is credited with three of the defining principles of the science of stratigraphy: the law of superposition: "...at the time when any given stratum was being formed, all the matter resting upon it was fluid, and, therefore, at the time when the lower stratum was being formed, none of the upper strata existed"; the principle of original horizontality: "Strata either perpendicular to the horizon or inclined to the horizon were at one time parallel to the horizon"; the principle of lateral continuity: "Material forming any stratum were continuous over the surface of the Earth unless some other solid bodies stood in the way"; and the principle of cross-cutting discontinuities: "If a body or discontinuity cuts across a stratum, it must have formed after that stratum." Another principle, known simply as Steno's law, or Steno's law of constant angles, states that the angles between corresponding faces on crystals are the same for all specimens of the same mineral, a fundamental breakthrough that formed the basis of all subsequent inquiries into crystal structure. [source: Wikipedia bio] For more on Steno's religious views and relation to his scientific endeavors, see: Frank Sobiech, "Heart, God, Cross: The Spirituality of the Anatomist, Geologist, and Bishop Dr. Nicholas Steno."

**Ole Christensen Rømer** (1644-1710) In 1676 he made the first quantitative measurements of the speed of light. Rømer didn't give a value for the speed of light but many others calculated a speed from his data, the first being Christiaan Huygens. Rømer also invented the Meridian circle, the Altazimuth and the Passage Instrument. [source: Wikipedia bio]

**John Flamsteed** (1646-1719) He was the key figure in the founding of the Royal Greenwich Observatory. Flamsteed accurately calculated the solar eclipses of 1666 and 1668. He was responsible for several of the earliest recorded sightings of the planet Uranus (the first in 1690): he mistook it for a star. His work *Historia Coelestis Britannica* contained Flamsteed's observations, and included a catalogue of 2,935 stars to much greater accuracy than any prior work. [source: Wikipedia bio]

# Chapter Five

## 36 Catholic, Protestant and Otherwise Religious Prominent Scientists: 1700-1800 (From Newton to Linnaeus, Boscovich, and Lavoisier)

**Antonie Philips van Leeuwenhoek** (1632-1723) Commonly known as "the Father of Microbiology" and is also known for his work on the improvement of the microscope. Using his handcrafted microscopes he was the first to observe and describe single celled organisms: what we now refer to as microorganisms. He was also the first to record microscopic observations of muscle fibers, bacteria, spermatozoa and blood flow in capillaries (small blood vessels). He discovered bacteria, spermatozoa, and the banded pattern of muscular fibers. His determined that smaller organisms procreate just as larger organisms do, thus overturning the traditional belief of the time in the spontaneous generation of such organisms: a belief generally held by 17th century scientists. [source: Wikipedia bio]

**Isaac Newton** (1643-1727; Arian theist) His 1687 publication of the *Philosophiæ Naturalis Principia Mathematica* (usually called the *Principia*) is considered to be among the most influential books in the history of science, laying the groundwork for most of classical mechanics. In this work, Newton described universal gravitation and the three laws of motion which dominated the scientific view of the physical universe for the next three centuries. Newton showed that the motions of objects on Earth and of celestial bodies are governed by the same set of natural laws by demonstrating the consistency between Kepler's laws of planetary motion and his theory of gravitation, thus removing the last doubts about heliocentrism and advancing the scientific revolution. Newton built the first practical reflecting telescope (Newtonian telescope) and developed a theory of color based on the observation that a prism decomposes white light into the many colours that form the visible spectrum (later expanded into his *Opticks*). He also formulated an empirical law of cooling and studied the speed of sound. In mathematics, Newton shares the credit with Gottfried Leibniz for the development of the differential and integral calculus. Newton is generally credited with the generalised binomial theorem, valid for any exponent. He discovered Newton's identities, Newton's method, classified cubic plane curves (polynomials of degree three in two variables), made substantial contributions to the theory of finite differences, and was the first to use fractional indices and to employ coordinate geometry to derive solutions to Diophantine equations. He approximated partial sums of the harmonic series by logarithms (a precursor to Euler's summation formula), and was the first to use power series with confidence and to revert power series. Newton remains influential to scientists, as demonstrated by a 2005 survey of members of Britain's Royal Society (formerly headed by Newton) asking who had the greater effect on the history of science, Newton or Albert Einstein. Royal Society scientists deemed Newton to have made the greater overall contribution. In 1999, an opinion poll of 100 of today's leading physicists voted Einstein the "greatest physicist ever;" with Newton the runner-up, while a parallel survey of rank-and-file physicists by the site PhysicsWeb gave the top spot to Newton. [source: Wikipedia bio] Newton wrote more than a million words on the Bible and theological topics, more than on science. He wrote in his famous book *Principia mathematica*:

> This most beautiful system of the sun, planets, and comets, could only proceed from the counsel and dominion of an intelligent and powerful Being . . . This Being governs all things, not as the soul of the world, but as the Lord over all. [source: Tripp]

Elsewhere, he wrote:

> Atheism is so senseless. When I look at the solar system, I see the earth at the right distance from the sun to receive the proper amounts of heat and light. This did not happen by chance. [source: Dimitrov]

For an in-depth treatment of the relationship of Newton's religious and scientific views, see: "The Complexity of Newton," by Steve Nakoneshny. For an exposition of the details of his heterodox theological positions, see: "Newton Reconsidered," by Stephen David Snobelen, and "Isaac Newton's Occult Studies" (Wikipedia).

**Gottfried Wilhelm Leibniz** (1646-1716) Leibniz occupies a grand place in both the history of philosophy and the history of mathematics. He invented infinitesimal calculus independently of Newton, and his notation has been in general use since then. He also invented the binary system, the foundation of virtually all modern computer architectures. Leibniz also made major contributions to physics and technology, and anticipated notions that surfaced much later in biology, medicine, geology, probability theory, psychology, linguistics, and information science. He also wrote on politics, law, ethics, theology, history, philosophy and philology. Leibniz saw that the uniqueness of prime factorization suggests a central role for prime numbers in the universal characteristic, a striking anticipation of Gödel numbering. Leibniz is the most important logician between Aristotle and 1847, when George Boole and Augustus De Morgan each published books that began modern formal logic. Leibniz enunciated the principal properties of what we now call conjunction, disjunction, negation, identity, set inclusion, and the empty set. Bertrand Russell went so far as to claim that Leibniz had developed logic in his unpublished writings to a level which was reached only 200 years later. Analytic and contemporary philosophy continue to invoke his notions of identity, individuation, and possible worlds. Although the mathematical notion of function was implicit in trigonometric and logarithmic tables, which existed in his day, Leibniz was the first, in 1692 and 1694, to employ it explicitly, to denote any of several geometric concepts derived from a curve, such as abscissa, ordinate, tangent, chord, and the perpendicular. Leibniz was the first to see that the coefficients of a system of linear equations could be arranged into an array, now called a matrix, which can be manipulated to find the solution of the system, if any. This method was later called Gaussian elimination. Leibniz discovered Boolean algebra and symbolic logic, also relevant to mathematics. Leibniz is credited, along with Sir Isaac Newton, with the inventing of infinitesimal calculus. He devised a new theory of motion (dynamics) based on kinetic energy and potential energy, which posited space as relative, whereas Newton felt strongly space was absolute. He anticipated Albert Einstein by arguing, against Newton, that space, time and motion are relative, not absolute. Leibniz's rule is an important, if often overlooked, step in many proofs in diverse fields of physics. The principle of sufficient reason has been invoked in recent cosmology, and his identity of indiscernibles in quantum mechanics, a field some even credit him with having anticipated in some sense. Those who advocate digital philosophy, a recent direction in cosmology, claim Leibniz as a precursor. By proposing that the earth has a molten core, he anticipated modern geology. In embryology, he was a preformationist, but also proposed that organisms are the outcome of a combination of an infinite number of possible microstructures and of their powers. In the life sciences and paleontology, he revealed an amazing transformist intuition, fueled by his study of comparative anatomy and fossils. In psychology, he anticipated the distinction between conscious and unconscious states. He even proposed something akin to what much later emerged as game theory. In sociology he laid the ground for communication theory. He urged that theory be combined with practical application, and thus has been claimed as the father of applied science. He designed wind-driven propellers and water pumps, mining machines to extract ore, hydraulic presses, lamps, submarines, clocks, etc. With Denis Papin, he invented a steam engine. He even proposed a method for desalinating water. Leibniz may have been the first computer scientist and information theorist. Early in life, he documented the binary number system (base 2), which is used on computers, then revisited that system throughout his career. He anticipated Lagrangian interpolation and algorithmic information theory. His calculus ratiocinator anticipated aspects of the universal Turing machine. Leibniz was one of the founders of library science. [source: Wikipedia bio] See a collection of extensive papers from a conference at Notre Dame University: examining Leibniz's highly influential theodicy, or defense of God against the charge that He is responsible for evil. For a treatment of his visionary ecumenism, see Markku Roinila, "The Reunion of the Churches." For a book-length treatment of his religious views, see: Maria Rosa Antognazza, *Leibniz on the Trinity and the Incarnation: Reason and Revelation in the Seventeenth Century* (Yale Univ. Press, 2008).

**Pierre Varignon** (1654-1722; Jesuit priest) His principal contributions were to graphic statics and mechanics. He studied the application of differential calculus to fluid flow and to water clocks and devised a mechanical explanation of gravitation. He also applied calculus to spring-driven clocks. [source: Wikipedia bio]

**John Woodward** (1665-1728) He showed that the stony surface of the earth was divided into strata, and that the enclosed shells were originally generated at sea, and in his elaborate *Catalogue* he described his rocks, minerals and fossils in a manner far in advance of the age. He was a founder of experimental plant-physiology, for he was one of the first to employ the method of water-culture, and to make refined experiments for the investigation of plant life. [source: Wikipedia bio]

**Francis Hauksbee** (1666-1713) He is best known for his work on electricity and electrostatic repulsion. By 1705, Hauksbee had discovered that if he placed a small amount of mercury in the glass of his modified version of Otto von Guericke's generator, evacuated the air from it and built up a charge on the ball, a glow was visible if he placed his hand on the outside of the ball. This effect later became the basis of the gas-discharge lamp, which led to neon lighting and mercury vapor lamps. [source: Wikipedia bio]

**Giovanni Girolamo Saccheri** (1667-1733; Jesuit priest) Primarily known today for his last publication of *Euclid Freed of Every Flaw*: the first European work of non-Euclidean geometry. Many of Saccheri's ideas have precedent in the 11th Century Persian Omar Khayyam's *Discussion of Difficulties in Euclid*. It is unclear whether Saccheri had access to this work in translation, or developed his ideas independently. [source: Wikipedia bio]

**Stephen Hales** (1677-1761) Demonstrated that plants absorb air, discovered the dangers of breathing stale air, and invented a ventilator which improved survival rates when employed on ships, in hospitals and in prisons. Hales is also credited with important work in pneumatic chemistry. He was the first to determine arterial blood pressure, was a pioneer of experimental physiology, and also invented the surgical forceps. [source: Wikipedia bio] He was an Anglican curate who worked with the Society for the Promotion of Christian Knowledge.

**James Bradley** (1693-1762) Best known for two fundamental discoveries in astronomy: the aberration of light and the nutation of the Earth's axis. The former discovery was, for all realistic purposes, conclusive evidence for the movement of the Earth. The theory of the aberration also gave Bradley a means to improve on the accuracy of the previous estimate of the speed of light, which had previously been shown to be finite by the work of Ole Rømer and others. [source: Wikipedia bio]

**William Smellie** (1697-1763) Preeminent obstetrician of his time and called *the father of British midwifery*. He described the mechanism of labor, designed obstetrical forceps, and devised a maneuver to deliver the head of a breech. [source: Wikipedia bio]

**Charles François de Cisternay du Fay** (1698-1739) Discovered the existence of two types of electricity and named them "vitreous" and "resinous" (later known as positive and negative charge respectively.) He noted the difference between conductors and insulators, calling them 'electrics' and 'non-electrics' for their ability to produce contact electrification. He also discovered that alike-charged objects would repel each other and that unlike-charged objects attract. [source: Wikipedia bio]

**John Bartram** (1699-1777) American (Quaker) botanist and horticulturalist. Carolus Linnaeus said he was the "greatest natural botanist in the world." Has been called the "father of American Botany", and was one of the first practicing Linnaean botanists in North America. His plant specimens were forwarded to Linnaeus, Dillenius and Gronovius. [source: Wikipedia bio]

**Jean-Antoine Nollet** (1700-1770; priest) The first to recognize the importance of sharp points on the conductors in the discharge of electricity. This was later applied practically in the construction of the lightning-rod. He also studied the conduction of electricity in tubes, in smoke, vapours, steam, the influence of electric charges on evaporation, vegetation, and animal life. His discovery in 1748 of the osmosis of water through a bladder into alcohol was the starting-point of that branch of physics. [source: *Catholic Encyclopedia* bio]

**Daniel Bernoulli** (1700-1782) He is particularly remembered for his applications of mathematics to mechanics, especially fluid mechanics, and for his pioneering work in probability and statistics. Bernoulli's work is still studied at length by many schools of science throughout the world. He pointed out for the first time the frequent desirability of resolving a compound motion into motions of translation and motions of rotation, and wrote a large number of papers on various mechanical questions, especially on problems connected with vibrating strings. He authored the 1738 work, *Specimen theoriae novae de mensura sortis (Exposition of a New Theory on the Measurement of Risk)*, in which the St. Petersburg paradox was the base of the economic theory of risk aversion, risk premium and utility. He is the earliest writer who attempted to formulate a kinetic theory of gases, and he applied the idea to explain Boyle's law. He worked with his good friend Leonhard Eule on elasticity and the development of the Euler-Bernoulli beam equation. Bernoulli's principle is of critical use in aerodynamics. [source: Wikipedia bio]

**Thomas Bayes** (c. 1702-1761) Mathematician and Presbyterian minister, known for having formulated a specific case of the theorem that bears his name: Bayes' theorem. In his *An Introduction to the Doctrine of Fluxions, and a Defence of the Mathematicians Against the Objections of the Author of the Analyst* (1736) he defended the logical foundation of Isaac Newton's calculus ("fluxions") against the criticism of George Berkeley. Bayesian probability is the name given to several related interpretations of probability, which have in common the notion of probability as something like a partial belief, rather than a frequency. This allows the application of probability to all sorts of propositions rather than just ones that come with a reference class. "Bayesian" has been used in this sense since about 1950. [source: Wikipedia bio]

**Benjamin Franklin** (1706-1790) Invented the lightning rod, bifocals, the Franklin stove, a carriage odometer, the glass 'armonica', and the flexible urinary catheter. He was the first to chart and codify the Gulf Stream in the Atlantic Ocean. Franklin proposed that "vitreous" and "resinous" electricity were not different types of "electrical fluid" (as electricity was called then), but the same electrical fluid under different pressures. He was the first to label them as positive and negative respectively, and he was the first to discover the principle of conservation of charge. He understood the concept of electrical ground. Franklin was, along with his contemporary Leonard Euler, the only major scientist who supported Christiaan Huygens' wave theory of light, which was basically ignored by the rest of the scientific community. In the 18th century Newton's corpuscular theory was held to be true; only after the famous Young's slit experiment were most scientists persuaded to believe Huygens' theory. He deduced that storms do not always travel in the direction of the prevailing wind, a concept that would have great influence in meteorology. Franklin was not an atheist, as some have erroneously claimed. In his *Autobiography*, he wrote:

> I never was without some religious principles. I never doubted, for instance, the existence of the Deity; that He made the world, and governed it by His providence; that the most acceptable service of God was the doing good to man; that our souls are immortal; and that all crime will be punished, and virtue rewarded, either here or hereafter. [source: Wikipedia bio]

**Vincent Riccati** (1707-1775; Jesuit priest) He worked together with Girolamo Saladini in publishing his discovery, the hyperbolic functions -- although Lambert is often incorrectly given this credit. Riccati not only introduced these new functions, but also derived the integral formulas connected with them. He then went on to derive the integral formulas for the trigonometric functions. His book *Institutiones* is recognized as the first extensive treatise on integral calculus. [source: *Adventures of Early Jesuit Scientists* bio]

**Carl Linnaeus** (1707-1778) Botanist, physician, and zoologist, who laid the foundations for the modern scheme (following Gaspard Bauhin and Johann Bauhin) of binomial nomenclature. He is known as the father of modern taxonomy, and is also considered one of the fathers of modern ecology. The establishment of universally accepted conventions for the naming of organisms was Linnaeus' main contribution to taxonomy. In addition Linnaeus developed what became known as the *Linnaean taxonomy*; the system of scientific classification now widely used in the biological sciences. [source: Wikipedia bio] Linnaeus stated: "The Earth's creation is the glory of God, as seen from the works of Nature by Man alone." One biographical page observed: "Linnaeus regarded himself as a man of the Enlightenment and as a traditional Christian. He had a rational view of the economic utility of science, but at the same time he nurtured an almost religious feeling for the beauty of nature and the magnificence of Creation."

**Leonhard Euler** (1707-1783) Introduced and popularized several notational conventions through his numerous and widely circulated textbooks. Most notably, he introduced the concept of a function and was the first to write $f(x)$ to denote the function $f$ applied to the argument $x$. He also introduced the modern notation for the trigonometric functions, the letter $e$ for the base of the natural logarithm (now also known as Euler's number), the Greek letter $\Sigma$ for summations and the letter $i$ to denote the imaginary unit. The use of the Greek letter $\pi$ to denote the ratio of a circle's circumference to its diameter was also popularized by Euler. While some of Euler's proofs are not acceptable by modern standards of mathematical rigour, his ideas led to many great advances. Euler is well-known in analysis for his frequent use and development of power series, the expression of functions as sums of infinitely many terms. Notably, Euler directly proved the power series expansions for $e$ and the inverse tangent function. His daring use of power series enabled him to solve the famous Basel problem in 1735. He introduced the use of the exponential function and logarithms in analytic proofs. He discovered ways to express various logarithmic functions using power series, and he successfully defined logarithms for negative and complex numbers, thus greatly expanding the scope of mathematical applications of logarithms. He also defined the exponential function for complex numbers, and discovered its relation to the trigonometric functions. Euler's identity was called "the most remarkable formula in mathematics" by Richard Feynman. Euler elaborated the theory of higher transcendental functions by introducing the gamma function and introduced a new method for solving quartic equations. He also found a way to calculate integrals with complex limits, foreshadowing the development of modern complex analysis, and invented the calculus of variations including its best-known result, the Euler–Lagrange equation. Euler also pioneered the use of analytic methods to solve number theory problems. In doing so, he united two disparate branches of mathematics and introduced a new field of study, analytic number theory. In breaking ground for this new field, Euler created the theory of hypergeometric series, q-series, hyperbolic trigonometric functions and the analytic theory of continued fractions. For example, he proved the infinitude of primes using the divergence of the harmonic series, and he used analytic methods to gain some understanding of the way prime numbers are distributed. Euler's work in this area led to the development of the prime number theorem. He also invented the totient function $\varphi(n)$ which is the number of positive integers less than or equal to the integer $n$ that are coprime to $n$. He contributed significantly to the theory of perfect numbers. He integrated Leibniz's differential calculus with Newton's Method of Fluxions, and developed tools that made it easier to apply calculus to physical problems. He made great strides in improving the numerical approximation of integrals, inventing what are now known as the Euler approximations. The most notable of these approximations are Euler's method and the Euler–Maclaurin formula. He also facilitated the use of differential equations, in particular introducing the Euler–Mascheroni constant. Euler helped develop the Euler–Bernoulli beam equation, which became a cornerstone of engineering. Aside from successfully applying his analytic tools to problems in classical mechanics, Euler also applied these techniques to celestial problems. His accomplishments include determining with great accuracy the orbits of comets and other celestial bodies, understanding the nature of comets, and calculating the parallax of the sun. His calculations also contributed to the development of accurate longitude tables. His 1740s papers on optics helped ensure that the wave theory of light proposed by Christian Huygens would become the dominant mode of thought, at least until the development of the quantum theory of light. He was a devout Christian (and believer in biblical inerrancy) who wrote apologetics and argued forcefully against the prominent atheists of his time. [source: Wikipedia bio]

**Georges-Louis Leclerc, Comte de Buffon** (1707-1788) He developed a concept of the "unity of type," a precursor of comparative anatomy. More than anyone else, he was responsible for the acceptance of a long-time scale for the history of the earth. Buffon noted that despite similar environments, different regions have distinct plants and animals, a concept later known as Buffon's Law, widely considered the first principle of biogeography. [source: Wikipedia bio]

**Mikhail Lomonosov** (1711-1765) He determined that the commonly accepted phlogiston theory was false. He regarded heat as a form of motion, suggested the wave theory of light, contributed to the formulation of the kinetic theory of gases, and stated the idea of conservation of matter. Lomonosov was the first person to record the freezing of mercury and to hypothesize the existence of an atmosphere on Venus. He demonstrated the organic origin of soil, peat, coal, petroleum and amber. In 1745, he published a catalogue of over 3,000 minerals, and in 1760, he explained the formation of icebergs. As a geographer, Lomonosov got close to the theory of continental drift, theoretically predicted the existence of Antarctica (he argued that icebergs of the South Ocean could only be formed on a dry land covered with ice). [source: Wikipedia bio]

**Roger Joseph Boscovich** (1711-1787; Jesuit priest) He is famous for his atomic theory and made many important contributions to astronomy, including the first geometric procedure for determining the equator of a rotating planet and for computing the orbit of a planet. In 1753 he also discovered the absence of atmosphere on the Moon. His atomic theory, given as a clear, precisely-formulated system utilizing principles of Newtonian mechanics inspired Michael Faraday to develop field theory for electromagnetic interaction. Some even claim that Boscovichian atomism was a basis for Albert Einstein's attempts for a unified field theory and that he was the first to envisage, seek, and propose a mathematical theory of all the forces of Nature; the first scientific theory of everything. [source: Wikipedia bio] Boscovich thought that atoms contain smaller parts, which in turn contain still smaller parts, and so forth down to the fundamental building blocks of matter. He felt that these building blocks must be geometric points with no size at all. Today, most atomic physicists accept a modern form of this idea. [source] Boscovich developed the idea that all phenomena arise from the spatial patterns of identical point particles (*puncta*) interacting in pairs according to an oscillatory law that determines their relative acceleration. This view of matter is akin to that of recent physics in that it is relational, structural, and kinematic. It contains three original features: (1) Material permanence without spatial extension; (2) Spatial relations without absolute space; and (3) Kinematic action without Newtonian forces. Boscovich helped emancipate physics from naive atomism's uncritical assumption that the ultimate units of matter are small, individual, rigid pieces possessing shape, size, weight, and other properties. [source]

**William Hunter** (1718-1783) Lleading teacher of anatomy, and the outstanding obstetrician of his day. To orthopaedic surgeons he is famous for his studies on bone and cartilage. [source: Wikipedia bio]

**Maria Gaetana Agnesi** (1718-1799; nun) She is credited with writing the first book discussing both differential and integral calculus. According to Dirk Jan Struik, Agnesi is "the first important woman mathematician since Hypatia (fifth century A. D.)". By her twentieth year because she strongly desired to enter a convent. Although her father refused to grant this wish, he agreed to let her live from that time on in an almost conventual semi-retirement, avoiding all interactions with society and devoting herself entirely to the study of mathematics. The most valuable result of her labours was the *Instituzioni analitiche ad uso della gioventù italiana*, a work of great merit, which was published in 1748 and "was regarded as the best introduction extant to the works of Euler." The first volume treats of the analysis of finite quantities and the second of the analysis of infinitesimals. In 1750, on the illness of her father, she was appointed by Pope Benedict XIV to the chair of mathematics and natural philosophy at Bologna. She was the first woman to be appointed professor at a university. After the death of her father in 1752 she carried out a long-cherished purpose by giving herself to the study of theology, and especially of the Fathers and devoted herself to the poor, homeless, and sick. After holding for some years the office of directress of the Hospice Trivulzio for Blue Nuns at Milan, she herself joined the sisterhood, and in this austere order ended her days. [source: Wikipedia bio]

**Christian Mayer** (1719-1783; Jesuit priest) He is most noted for pioneering the study of binary stars, although his equipment was ill-suitable for distinguishing between true binaries and coincident star alignments. In 1777-78 he compiled a catalog of 80 double stars. [source: Wikipedia bio]

**John Michell** (1724-1793) He was an Anglican priest whose scientific work spanned a wide range of subjects from astronomy to geology, optics, and gravitation. Michell conceived, sometime before 1783, the experiment now known as the Cavendish experiment. It was the first to measure the force of gravity between masses in the laboratory and produced the first accurate values for the mass of the Earth and the gravitational constant. He wrote a lucid exposition of the nature of magnetic induction. His most important geological essay was entitled "Conjectures concerning the Cause and Observations upon the Phaenomena of Earthquakes" (*Philosophical Transactions*, li. 1760), which showed a remarkable knowledge of geological strata. He was thus one of the founders of seismology. More recently, Michell's main "claim to fame" is considered to be his letter to Cavendish, published in 1784, on the effect of gravity on light. This paper was only generally "rediscovered" in the 1970s and is now recognised as anticipating several astronomical ideas that had been considered to be 20th century innovations. Michell is now credited with being the first to study the case of a heavenly object massive enough to prevent light from escaping (the concept of escape velocity was well known at the time). Such an object, which he called a "dark star" (the predecessor of the modern idea of a black hole under general relativity) would not be directly visible, but could be identified by the motions of a companion star if it was part of a binary system. Michell also derived the radius for such an object based on its mass, which corresponds roughly to what is called the Schwarzschild Radius in general relativity. Michell also suggested using a prism to measure the gravitational weakening of starlight due to the surface gravity of the source ("gravitational shift"). Michell acknowledged that some of these ideas were not technically practical at the time, but wrote that he hoped they would be useful to future generations. By the time that Michell's paper was "resurrected" nearly two centuries later, these ideas had been reinvented by others. [source: Wikipedia bio]

**James Hutton** (1726-1797; deist) He is considered the father of modern geology. His theories of geology and geologic time, also called deep time, came to be included in theories which were called plutonism and uniformitarianism. In 1785, Hutton found granite penetrating metamorphic schists, in a way that indicated that the granite had been molten at the time. This showed to him that granite formed from cooling of molten rock, not precipitation out of water as others at the time believed, and that the granite must be younger than the schists. Hutton proposed that the interior of the Earth was hot, and that this heat was the engine which drove the creation of new rock: land was eroded by air and water and deposited as layers in the sea; heat then consolidated the sediment into stone, and uplifted it into new lands. This theory was dubbed "Plutonist" in contrast to the flood-oriented theory. Rather than accepting that the earth was no more than a few thousand years old, he maintained that the Earth must be much older, with a history extending indefinitely into the distant past. His main line of argument was that the tremendous displacements and changes he was seeing did not happen in a short period of time by means of catastrophe, but that processes still happening on the Earth in the present day had caused them. As these processes were very gradual, the Earth needed to be ancient, in order to allow time for the changes. He investigated the available data regarding rainfall and climate in different regions of the globe, and came to the conclusion that the rainfall is regulated by the humidity of the air on the one hand, and mixing of different air currents in the higher atmosphere on the other. Hutton also advocated uniformitarianism for living creatures too – evolution, in a sense – and even suggested natural selection as a possible mechanism affecting them. Hutton saw his "principle of variation" as explaining the development of varieties, but rejected the idea of evolution originating species as a "romantic fantasy." [source: Wikipedia bio]

**John Hunter** (1728-1793) He was an early advocate of careful observation and scientific method in medicine. Among his numerous contributions to medical science are:

- study of human teeth
- extensive study of inflammation
- an understanding of the nature of digestion, and verifying that fats are absorbed into the lacteals, a type of small intestine lymphatic capillary, and not into the intestinal blood capillaries as was generally accepted.
- the first complete study of the development of a child
- proof that the maternal and foetal blood supplies are separate
- unravelling of one of the major anatomical mysteries of the time – the role of the lymphatic system [source: Wikipedia bio]

**Lazzaro Spallanzani** (1729-1799; priest) His life was one of incessant eager questioning of nature on all sides, and he was an original genius, capable of stating and solving problems in all departments of science -- at one time finding the true explanation of stone skipping (formerly attributed to the elasticity of water) and at another helping to lay the foundations of our modern vulcanology and meteorology. He researched the theory about the spontaneous generation of cellular life and suggested that microbes move through the air and that they could be killed through boiling, paving the way for later research by Louis Pasteur, who defeated the theory of spontaneous generation. He also discovered and described animal (mammal) reproduction, showing that it requires both semen and an ovum. He is also famous for extensive experiments on the navigation in complete darkness by bats, where he concluded that bats use sound and their ears for navigation in total darkness (see animal echolocation). He first interpreted the process of digestion, which he connected to chemical solution, taking place primarily in the stomach, by the action of the gastric juice. [source: Wikipedia bio] His studies in regeneration are still classic. He showed experimentally that many animals like the lizard and the snail, if accidentally injured, regenerate important parts of their bodies; the land snail regenerates even its head. It was afterwards shown that this does not contain the brain, but it does contain eyes, mouth, tongue, and teeth, and these are all regenerated. [*Catholic Encyclopedia* bio]

**Joseph Priestley** (1733-1804) He is usually credited with the discovery of oxygen, having isolated it in its gaseous state. During his lifetime, Priestley's considerable scientific reputation rested on his invention of soda water, his writings on electricity, and his discovery of several "airs" (gases): nitric oxide (NO), anhydrous hydrochloric acid (HCl), ammonia ($NH_3$), and nitrous oxide ($N_2O$). His experiments would eventually lead to the discovery of photosynthesis. [source: Wikipedia bio]

**Charles-Augustin de Coulomb** (1736-1806) He is best known for developing Coulomb's law, the definition of the electrostatic force of attraction and repulsion. He was a pioneer in the field of geotechnical engineering for his contribution to retaining wall design. He successfully experimented on the torsional force for metal wires and different forms of his torsion balance. He used the instrument with great success for the experimental investigation of the distribution of charge on surfaces, of the laws of electrical and magnetic force and of the mathematical theory of which he may also be regarded as the founder. In his *Deuxieme Mémoire sur l'Electricité et le Magnétisme* he made a "determination according to which laws both the Magnetic and the Electric fluids act, either by repulsion or by attraction." Coulomb explained, among many other related things, the laws of attraction and repulsion between electric charges and magnetic poles, although he did not find any relationship between the two phenomena. He thought that the attraction and repulsion were due to different kinds of fluids. [source: Wikipedia bio]

**Luigi Galvani** (1737-1798) In 1771, he discovered that the muscles of dead frogs legs twitched when struck by a spark. This was one of the first forays into the study of bioelectricity, a field that studies the electrical patterns and signals of the nervous system. Galvani was the first investigator to appreciate the relationship between electricity and animation — or life. This finding provided the basis for the current understanding that electrical energy (carried by ions), and not air or fluid as in earlier balloonist theories, is the impetus behind muscle movement. [source: Wikipedia bio]

**Frederick William Herschel** (1738-1822) He discovered the planet Uranus in addition to several of its major moons such as Titania and Oberon, and two moons of Saturn, Mimas and Enceladus; as well as infrared radiation. During the course of his career, he constructed more than four hundred telescopes and designed what is now known as the Herschelian telescope. Herschel discovered that unfilled telescope apertures can be used to obtain high angular resolution, something which became the essential basis for interferometric imaging in astronomy (in particular Aperture Masking Interferometry and hypertelescopes). He worked on creating an extensive catalogue of nebulae. He continued to work on double stars, and was the first to discover that most double stars are not mere optical doubles as had been supposed previously, but are true binary stars, thus providing the first evidence that Newton's laws of gravitation apply outside the solar system. He also measured the axial tilt of the planet Mars and discovered that the martian ice caps changed size with the planet's seasons. From studying the proper motion of stars, he was the first to realize that the solar system is moving through space, and he determined the approximate direction of that movement. He also studied the structure of the Milky Way and concluded that it was in the shape of a disk. He also coined the word "asteroid", meaning *star-like*. [source: Wikipedia bio]

**Juan Molina** (1740-1829; Jesuit priest) Juan described an analogy between living organisms and minerals. He proposed an idea of the gradual evolution of human beings, thereby anticipating Darwin's theory of evolution. In an 1815 work on nature's three kingdoms (mineral, vegetable and animal) he describes the Creator's plan for a continuous seamless chain of life from mineral life to vegetable life to animal life with no discrete discontinuous steps. Crystalline minerals tend to gather together in preparation for the higher form of vegetable life which then evolve into animal life. [source: *Adventures of Early Jesuit Scientists* bio]

**Antoine Lavoisier** (1743-1794) The "father of modern chemistry"; he also stated the first version of the law of conservation of mass, recognized and named oxygen and hydrogen, abolished the phlogiston theory, helped construct the metric system, wrote the first extensive list of elements, and helped to reform chemical nomenclature. He discovered that, although matter may change its form or shape, its mass always remains the same. Lavoisier demonstrated the role of oxygen in the rusting of metal, as well as oxygen's role in animal and plant respiration. Lavoisier's researches included some of the first truly quantitative chemical experiments. He carefully weighed the reactants and products in a chemical reaction, which was a crucial step in the advancement of chemistry. He investigated the composition of water and air, which at the time were considered elements. He determined that the components of water were oxygen and hydrogen, and that air was a mixture of gases, primarily nitrogen and oxygen. His *Traité Élémentaire de Chimie* (*Elementary Treatise on Chemistry*, 1789) is considered to be the first modern chemistry textbook. This text clarified the concept of an element as a substance that could not be broken down by any known method of chemical analysis, and presented Lavoisier's theory of the formation of chemical compounds from elements. He established the consistent use of the chemical balance. Lavoisier also contributed to early ideas on composition and chemical changes by stating the radical theory, believing that radicals, which function as a single group in a chemical process, combine with oxygen in reactions. He also introduced the possibility of [allotropy in chemical elements](#) when he discovered that diamond is a crystalline form of carbon. Overall, his contributions are considered the most important in advancing chemistry to the level reached in physics and mathematics during the 18th century. [source: [Wikipedia bio](#)]

**Marie François Xavier Bichat** (1771-1802) Anatomist and physiologist, best remembered as the father of modern [histology](#) and [pathology](#). Despite the fact that he worked without a microscope he was able to advance greatly the understanding of the human body. He was the first to introduce the notion of *tissue* (tissues) as distinct entities. He maintained that diseases attacked tissues rather than whole organs. His work, *Anatomie générale* (1801) contains the fruits of his most profound and original researches. [source: [Wikipedia bio](#)]

# Chapter Six

## 41 Catholic, Protestant and Otherwise Religious Prominent Scientists: 1800-1850 (From Dalton to Humboldt, Cuvier, and Faraday)

Alessandro Volta (1745-1827) Physicist known especially for the development of the first electric cell in 1800. In 1776-77 Volta studied the chemistry of gases. He discovered methane by collecting the gas from marshes. He devised experiments such as the ignition of methane by an electric spark in a closed vessel. Volta also studied what we now call electrical capacitance, developing separate means to study both electrical potential (V) and charge (Q), and discovering that for a given object they are proportional. This may be called Volta's Law of capacitance, and likely for this work the unit of electrical potential has been named the Volt. He discovered the electrochemical series, and the law that the electromotive force (emf) of a galvanic cell, consisting of a pair of metal electrodes separated by electrolyte, is the difference between their two electrode potentials. (Thus, two identical electrodes and a common electrolyte give zero net emf.) This may be called Volta's Law of the electrochemical series. In 1800 he invented the voltaic pile, an early electric battery, which produced a steady electric current. Volta had determined that the most effective pair of dissimilar metals to produce electricity was zinc and silver. The battery made by Volta is credited as the first electrochemical cell. [source: Wikipedia bio] Three practical units have been named after Catholic electrical pioneers; the volt, the unit of electrical pressure, in honour of Volta; the coulomb, the unit of electrical quantity, in honour of Charles Augustin de Coulomb; and the ampere, the unit of current, in honour of André-Marie Ampère. [source: <u>Catholic Encyclopedia bio</u>]

**Giuseppe Piazzi** (1746-1826; priest) He supervised the compilation of the Palermo Catalogue of stars, containing 7,646 star entries with unprecedented precision. He discovered <u>Ceres</u>, today known as the largest member of the <u>asteroid belt</u>. On January 1, 1801, Piazzi discovered a "stellar object" that moved against the background of stars. At first he thought it was a fixed star, but once he noticed that it moved, he became convinced it was a planet, or as he called it, "a new star". After its orbit was better determined, it was clear that Piazzi's assumption was correct and this object was not a comet but more like a small planet. [source: <u>Wikipedia bio</u>]

**John Playfair** (1748-1819) Scientist and mathematician. He is perhaps best known for his book *Illustrations of the Huttonian Theory of the Earth* (1802), which summarized the work of <u>James Hutton</u>. It was through this book that Hutton's principle of <u>uniformitarianism</u>, later taken up by <u>Charles Lyell</u>, first reached a wide audience. Playfair also is remembered for his proposal of an alternative to <u>Euclid</u>'s parallel postulate. His *<u>Elements of Geometry</u>* first appeared in 1795 and has passed through many editions. [source: <u>Wikipedia bio</u>]

**Edward Anthony Jenner** (1749-1823) He is widely credited as the pioneer of <u>smallpox vaccine</u>, and is sometimes referred to as the "Father of Immunology". His vaccine also laid the groundwork for modern-day discoveries in that field. [source: <u>Wikipedia bio</u>]

**Johann Wolfgang von Goethe** (1749-1832) Goethe wrote several works on plant <u>morphology</u>, and colour theory. His focus on morphology and what was later called <u>homology</u> influenced 19th century naturalists, though his ideas of transformation were about the continuing flux of living things and did not relate to contemporary ideas of "transformisme" or transmutation of species. Homology was used by Charles Darwin as strong evidence of common descent and of laws of variation. Goethe's studies led him to independently discover the human intermaxillary bone in 1784. He popularized the Goethe Barometer using a principle established by Toricelli. In 1810, Goethe published his *Theory of Colours*, which he considered his most important work. He was the first to systematically study the physiological effects of color. [source: Wikipedia bio]

**Heinrich Wilhelm Matthäus Olbers** (1758-1840) Devised the first satisfactory method of calculating cometary orbits and proposed that the asteroid belt was a remnant of a planet that had been destroyed. The current view of most scientists is that tidal effects from the planet Jupiter disrupt the formation of planets in the asteroid belt. Olbers' paradox, described by him in 1823, states that the darkness of the night sky conflicts with the supposition of an infinite and eternal static universe. [source: Wikipedia bio]

**William Kirby** (1759-1850) He is considered the "founder of entomology". Kirby produced his first major work, the *Monographia Apum Angliae* (*Monograph on the Bees of England*), in 1802. His purpose was both scientific and religious:

> The author of Scripture is also the author of Nature: and this visible world, by types indeed, and by symbols, declares the same truths as the Bible does by words. To make the naturalist a religious man – to turn his attention to the glory of God, that he may declare his works, and in the study of his creatures may see the loving-kindness of the Lord – may this in some measure be the fruit of my work… (*Correspondence*, 1800)

His *Introduction to Entomology*, a celebrated title, appeared in four volumes between 1815 and 1826. In 1830 he was invited to write one of the Bridgewater Treatises, his subject being *The*

*History, Habits, and Instincts of Animals* (2 vols., 1835). [source: Wikipedia bio]

**John Dalton** (1766-1844) Chemist, meteorologist and physicist, best known for his pioneering work in the development of modern atomic theory, and research into colour blindness. He studied the absorption of gases by water and other liquids and developed his law of partial pressures now known as Dalton's law. The idea of atoms arose in his mind as a purely physical concept, forced upon him by study of the physical properties of the atmosphere and other gases. The first published indications of this idea are to be found at the end of his paper on the absorption of gases already mentioned, which was read on 21 October 1803, though not published until 1805. Here he says:

> Why does not water admit its bulk of every kind of gas alike? This question I have duly considered, and though I am not able to satisfy myself completely I am nearly persuaded that the circumstance depends on the weight and number of the ultimate particles of the several gases.

Dalton proceeded to print his first published table of relative atomic weights. Assisted by the assumption that combination always takes place in the simplest possible way, he thus arrived at the idea that chemical combination takes place between particles of different weights. The extension of this idea to substances in general necessarily led him to the law of multiple proportions, and the comparison with experiment brilliantly confirmed his deduction. No evidence was then available to scientists to deduce how many atoms of each element combine to form compound molecules. But this or some other such rule was absolutely necessary to any incipient theory, since one needed an assumed molecular formula in order to calculate relative atomic weights. The idea that all atoms of a given element are identical in their physical and chemical properties is not precisely true, as we now know that different isotopes of an element have slightly varying weights. However, Dalton had created a theory of immense power and importance. Indeed, Dalton's innovation was fully as important for the future of the science as Antoine Laurent

Lavoisier's oxygen-based chemistry had been. He was a Quaker. [source: Wikipedia bio]

**Jean Baptiste Joseph Fourier** (1768-1830) Best known for initiating the investigation of Fourier series and their applications to problems of heat transfer and vibrations. Fourier also developed dimensional analysis, the method of representing physical units, such as velocity and acceleration, by their fundamental dimensions of mass, time, and length, to obtain relations between them. Fourier left an unfinished work on determinate equations that contained much original matter — in particular, there is a demonstration of Fourier's theorem on the position of the roots of an algebraic equation. [source: Wikipedia bio]

Georges Cuvier (1769-1832) Naturalist and zoologist, instrumental in establishing the fields of comparative anatomy and paleontology through his work in comparing living animals with fossils. He established, for the first time, the fact that African and Indian elephants were different species and that mammoths were not the same species as either African or Indian elephants and therefore must be extinct. At the time Cuvier presented his 1796 paper on living and fossil elephants, it was still widely believed that no species of animal had ever become extinct. Cuvier became an active proponent of the geological school of thought called catastrophism that maintained that many of the geological features of the earth and the past history of life could be explained by catastrophic events that had caused the extinction of many species of animals. Over the course of his career Cuvier came to believe that there had not been a single catastrophe but several, resulting in a succession of different faunas. The increasing interest in the topic of mass extinction starting in the late 20th century has led to a resurgence of interest among historians of science and other scholars in this aspect of Cuvier's work. Collaborating with Alexandre Brongniart, Cuvier identified characteristic fossils of different rock layers that they used to analyze the geological column, the ordered layers of sedimentary rock, of the Paris basin. They concluded that the layers had been laid down over an extended period during which there clearly had been faunal succession and that the area had been submerged under sea water at times and at other times under fresh water. Along with William Smith's work during the same period, the monograph helped establish the scientific discipline of stratigraphy. It was a major development in the history of paleontology and the history of geology. In 1800, Cuvier was the first to correctly identify in print, working only from a drawing, a fossil found in Bavaria as a small flying reptile, which he named the *Ptero-Dactyle* in 1809 (later Latinized as *Pterodactylus antiquus*)--the first known member of the diverse order of pterosaurs. Cuvier speculated that there had been a time when reptiles rather than mammals had been the dominant fauna. Cuvier published a long list of memoirs, partly relating to the bones of extinct animals, and partly detailing the results of observations on the skeletons of living animals, specially examined with a view of throwing light upon the structure and affinities of the fossil forms. The department of palaeontology dealing with the Mammalia may be said to have been essentially created and established by Cuvier. He was critical of the evolutionary theories proposed by his contemporaries Lamarck and Geoffroy Saint-Hilaire, which involved the gradual transmutation of one form into another. He repeatedly emphasized that his extensive experience with fossil material indicated that one fossil form does not, as a rule, gradually change into a succeeding, distinct fossil form. Instead, he said, the typical form makes an abrupt appearance in the fossil record, and persists unchanged to the time of its extinction (this is the well-documented paleontological phenomenon now referred to as "punctuated equilibrium"). In other words, Cuvier was a saltationist. While, like other saltationists, he offered no explanation of how saltational evolution might occur, he was skeptical of the gradual mechanisms of change that Lamarck and Geoffroy Saint-Hilaire proposed. He argued that one can only judge what a long time would produce by multiplying what a lesser time produces. Since a lesser time produced no organic changes, neither, probably, would a much longer time. He nowhere refers to the Bible in scientific argument. Nor did he advance the hypothesis of successive new creations. The harshness of his criticism and the strength of his reputation continued to discourage naturalists from speculating about the gradual transmutation of species, right up until Darwin published *On the Origin of Species* more than two decades after Cuvier's death. [source: Wikipedia bio]

**William Smith** (1769-1839) He is credited with creating the first nationwide geological map. Adam Sedgwick referred to Smith as "the Father of English Geology". As he observed the rock layers (or strata), he realised that they were arranged in a predictable pattern and that the various strata could always be found in the same relative positions. Additionally, each particular stratum could be identified by the fossils it contained, and the same succession of fossil groups from older to younger rocks could be found in many parts of England. He developed a testable hypothesis, which he termed The Principle of Faunal Succession. His nephew John Phillips lived with Smith during his youth and was his apprentice. He became a major figure in 19th century geology and paleontology—among other things he's credited as first to specify most of the table of geologic eras that is used today (1841). [source: Wikipedia bio]

**Alexander von Humboldt** (1769-1859) His quantitative work on botanical geography was foundational to the field of biogeography. Between 1799 and 1804, Humboldt traveled extensively in Latin America, exploring and describing it for the first time in a manner generally considered to be a modern scientific point of view. He was one of the first to propose that the lands bordering the Atlantic Ocean were once joined (South America and Africa in particular). Later, his five-volume work, *Kosmos* (1845), attempted to unify the various branches of scientific knowledge. Humboldt wrote "Nature herself is sublimely eloquent. The stars as they sparkle in firmament fill us with delight and ecstasy, and yet they all move in orbit marked out with mathematical precision." Charles Darwin stated: "He was the greatest traveling scientist who ever lived." [source: Wikipedia bio]

**Thomas Young** (1773-1829) Made notable scientific contributions to the fields of vision, light, solid mechanics, energy, and physiology. His contemporary Sir John Herschel called him a "truly original genius". Albert Einstein praised him in the 1931 foreword to an edition of Newton's *Opticks*. Other admirers include physicist Lord Rayleigh and Nobel laureate Philip Anderson. One of his most important achievements was to establish the wave theory of light. To support this viewpoint, he demonstrated with a ripple tank the idea of interference in the context of water waves. Young has also been called the founder of physiological optics. In 1793 he explained the mode in which the eye accommodates itself to vision at different distances as depending on change of the curvature of the crystalline lens; in 1801 he was the first to describe astigmatism; and in his *Lectures* he presented the hypothesis, afterwards developed by Hermann von Helmholtz, that colour perception depends on the presence in the retina of three kinds of nerve fibres which respond respectively to red, green and violet light. This foreshadowed the modern understanding of color vision, in particular the finding that the eye does indeed have three color receptors which are sensitive to different wavelength ranges. In 1804, Young developed the theory of capillary phenomena on the principle of surface tension. He also observed the constancy of the angle of contact of a liquid surface with a solid, and showed how from these two principles to deduce the phenomena of capillary action. He was the first to define the term "energy" in the modern sense. Young's equation describes the contact angle of a liquid drop on a plane solid surface as a function of the surface free energy, the interfacial free energy and the surface tension of the liquid. Young's equation was developed further some 60 years later by Dupré to account for thermodynamic effects, and this is known as the Young–Dupré equation. [source: Wikipedia bio]

**Sir George Cayley** (1773-1857) He was one of the most important people in the history of aeronautics. Sometimes called the "Father of Aviation", in 1799 he set forth the concept of the modern aeroplane as a fixed-wing flying machine with separate systems for lift, propulsion, and control. Often known as "the father of Aerodynamics", he was a pioneer of aeronautical engineering. Designer of the first successful glider to carry a human being aloft, he discovered and identified the four aerodynamic forces of flight—weight, lift, drag, and thrust—which are in effect on any flight vehicle. Modern aeroplane design is based on those discoveries including cambered wings. He is credited with the first major breakthrough in heavier-than-air flight. He designed the first actual model of an aeroplane and also diagrammed the elements of vertical flight. During some point prior to 1849 he designed and built a biplane powered with "flappers" in which an unknown ten-year-old boy flew. Later, he developed a larger scale glider (also probably fitted with "flappers") which flew across Brompton Dale in 1853. Among the many other things that he developed are self-righting lifeboats, tension-spoke wheels, the "Universal Railway" (his term for caterpillar tractors), automatic signals for railway crossings, seat belts, small scale helicopters, and a kind of prototypical internal combustion engine fueled by gunpowder. He also contributed in the fields of prosthetics, air engines, electricity, theatre architecture, ballistics, optics and land reclamation. [source: Wikipedia bio]

Sir Charles Bell (1774-1842) Anatomist, neurologist, and surgeon. In his *Essays on The Anatomy and Philosophy of Expression* (1824). In this work, he followed the principles of natural theology, asserting the existence of a uniquely human system of facial muscles in the service of a human species with a unique relationship to the Creator. His 1811 treatise, *An Idea of a New Anatomy of the Brain* described his experiments with animals, and is considered by many to be the founding stone of clinical neurology. Bell was one of the first physicians to combine the scientific study of neuroanatomy with clinical practice. In 1821, he described in the trajectory of the facial nerve and a disease, Bell's Palsy which led to the unilateral paralysis of facial muscles, in one of the classics of neurology, a paper to the Royal Society entitled *On the Nerves: Giving an Account of some Experiments on Their Structure an Functions, Which Lead to a New Arrangement of the System*. In 1833 he published the fourth

Bridgewater Treatise, *The Hand: Its Mechanism and Vital Endowments as Evincing Design.* [source: Wikipedia bio]
**John Pye-Smith** (1774-1851) Congregational theologian; associated with reconciling geological sciences with the Bible. He was elected to become the first Fellow of the Royal Society from a nonconformist background, and also elected a Fellow of the Geological Society at a time when there was considerable debate about accepting the idea of geological time, and if so to find ways of reconciling this with the teachings of the Old Testament. He authored *On the Relation Between the Holy Scriptures* and some parts of geological science in 1840. In that work he espoused the uniformitarian geology of Hutton and Lyell (themselves both Christians), and argued that it was not contradictory to the Christian faith. [source: Wikipedia bio]

**Jean-Baptiste Biot** (1774-1862) His work covers the entire field of experimental and mathematical physics, as well as ancient and modern astronomy. He was the champion of the corpuscular theory of light which he extended to some most ingenious explanations of the very complex phenomena of polarization. Biot discovered the laws of rotary polarization by crystalline bodies and applied these laws to the analysis of saccharine solutions. His fame rests chiefly on his work in polarization and double refraction of light. His Catholic religious views became more pronounced towards the end of his life. He is said to have received the sacrament of Confirmation at the hands of his own grandson. [source: *Catholic Encyclopedia* bio]

**André-Marie Ampère** (1775-1836) Physicist and mathematician who is generally regarded as one of the main discoverers of electromagnetism. On 11 September 1820 he heard of H. C. Ørsted's discovery that a magnetic needle is acted on by a voltaic current. Only a week later, on 18 September, Ampère presented a paper to the Academy containing a much more complete exposition of that and kindred phenomena. On the same day, Ampère also demonstrated before the Academy that parallel wires carrying currents attract or repel each other, depending on whether currents are in the same (attraction) or in opposite directions (repulsion). This laid the foundation of electrodynamics. The topic of electromagnetism thus begun, Ampère developed a mathematical theory which not only described the electromagnetic phenomena already observed, but also predicted many new ones. On the day of his wife's death in 1803 he wrote two verses from the Psalms, and the prayer, "O Lord, God of Mercy, unite me in Heaven with those whom you have permitted me to love on earth." Serious doubts harassed him at times, and made him very unhappy. Then he would take refuge in the reading of the Bible and the Fathers of the Church. [source: Wikipedia bio] His work alternated between mathematics, physics, and metaphysics. He published a number of articles on calculus, on curves, and other purely mathematical topics, as well as on chemistry and light, and even on zoölogy. In 1821 he suggested an electric telegraph, using separate wires for every letter. [source: *Catholic Encyclopedia* bio]

**Carl Friedrich Gauss** (1777-1855) Mathematician and scientist who contributed significantly to many fields, including number theory, statistics, analysis, differential geometry, geodesy, geophysics, electrostatics, astronomy and optics. Gauss had a remarkable influence in many fields of mathematics and science and is ranked as one of history's most influential mathematicians. He completed *Disquisitiones Arithmeticae*, his magnum opus, in 1798 at the age of 21. This work was fundamental in consolidating number theory as a discipline and has shaped the field to the present day. He invented modular arithmetic and became the first to prove the quadratic reciprocity law. He proved the fundamental theorem of algebra which states that every non-constant single-variable polynomial over the complex numbers has at least one root. His personal diaries indicate that he had made several important mathematical discoveries years or decades before his contemporaries published them. Mathematical historian Eric Temple Bell estimated that had Gauss timely published all of his discoveries, Gauss would have advanced mathematics by fifty years. He streamlined the cumbersome mathematics of 18th century orbital prediction, and his *Theory of Celestial Movement* remains a cornerstone of astronomical computation. It introduced the Gaussian gravitational constant, and contained an influential treatment of the method of least squares, a procedure used in all sciences to this day to minimize the impact of measurement error. Gauss was able to prove the method in 1809 under the assumption of normally distributed errors (see Gauss–Markov theorem; see also Gaussian). Gauss invented the heliotrope, an instrument that uses a mirror to reflect sunlight over great distances, to measure positions. He developed an interest in differential geometry, a field of mathematics dealing with curves and surfaces. Among other things he came up with the notion of Gaussian curvature. This led in 1828 to an important theorem, the Theorema Egregium (*remarkable theorem* in Latin), establishing an important property of the notion of curvature. In 1831 Gauss developed a fruitful collaboration with the physics professor Wilhelm Weber, leading to new knowledge in magnetism (including finding a representation for the unit of magnetism in terms of mass, length and time) and the discovery of Kirchhoff's circuit laws in electricity. They constructed the first electromechanical telegraph in 1833. He developed a method of measuring the horizontal intensity of the magnetic field of the earth which has been in use well into the second half of the 20th century and worked out the mathematical theory for separating the inner (core and crust) and outer (magnetospheric) sources of Earth's magnetic field. According to Dunnington, Gauss's religion was based upon the search for truth. He believed in "the immortality of the spiritual individuality, in a personal permanence after death, in a last order of things, in an eternal, righteous, omniscient and omnipotent God." [source: Wikipedia bio]

**Sir Humphry Davy** (1778-1829) He is probably best remembered today for his discoveries of several alkali and alkaline earth metals, as well as contributions to the discoveries of the elemental nature of chlorine and iodine. Berzelius called Davy's 1806 Bakerian Lecture *On Some Chemical Agencies of Electricity* "one of the best memoirs which has ever enriched the theory of chemistry." This paper was central to any chemical affinity theory in the first half of the nineteenth century. In 1815 he invented the Davy lamp, which allowed miners to work safely in the presence of flammable gases. As a boy, he was especially attracted by Bunyan's *Pilgrim's Progress*. Davy's lectures included spectacular and sometimes dangerous chemical demonstrations for his audience, a generous helping of references to divine creation, and genuine scientific information. Davy was a pioneer in the field of electrolysis using the voltaic pile to split up common compounds and thus prepare many new elements. He went on to electrolyse molten salts and discovered several new metals, especially sodium and potassium, highly reactive elements known as the alkali metals. He discovered potassium in 1807, calcium in 1808, and also magnesium, boron, and barium. He named chlorine in 1810 (it had been discovered in 1774 by Swedish chemist Carl Wilhelm Scheele). He studied laughing gas (nitrous oxide). He showed that iodine was an element. He proved (with Michael Faraday) that diamonds were composed of pure carbon. [source: Wikipedia bio]

**Benjamin Silliman** (1779-1864) He was a chemist and one of the first American professors of science (at Yale University). Silliman made a chemical analysis of the meteorite that fell near Weston, Connecticut: the first such scientific account in America and came to discover many of the constituent elements of many minerals. In 1854, he became the first person to fractionate petroleum by distillation. [source: Wikipedia bio] He was strongly Protestant. [source: *Christian Influences in the Sciences*, Daniel Graves]

**Bernard Bolzano** (1781-1848; priest) He was a mathematician, logician, philosopher, and theologian. Bolzano's posthumously published work *Paradoxien des Unendlichen (The Paradoxes of the Infinite)* was greatly admired by many of the eminent logicians who came after him, including Charles Sanders Peirce, Georg Cantor, and Richard Dedekind. Bolzano's main claim to fame, however, is his 1837 *Wissenschaftslehre* (*Theory of Science*), a work in four volumes that covered not only philosophy of science in the modern sense but also logic, epistemology and scientific pedagogy. The logical theory that

Bolzano developed in this work has come to be acknowledged as ground-breaking. His philosophical work was rediscovered by Edmund Husserl and Kazimierz Twardowski, both students of Franz Brentano. Through them, Bolzano became a formative influence on both phenomenology and analytic philosophy. He also did valuable work in mathematics, which remained virtually unknown until Otto Stolz rediscovered many of his lost journal articles and republished them in 1881. He gave the first purely analytic proof of the fundamental theorem of algebra, which had originally been proven by Gauss from geometrical considerations. He also gave the first purely analytic proof of the intermediate value theorem (also known as Bolzano's theorem). [source: Wikipedia bio]

**Sir David Brewster** (1781-1868) Physicist, mathematician, astronomer, and inventor. He initially studied theology and was licensed to preach. The most important subjects of his scientific inquiries can be enumerated under the following five headings:

>1.The laws of light polarization by reflection and refraction, and other quantitative laws of phenomena.
>2. The discovery of the polarizing structure induced by heat and pressure.
>3. The discovery of crystals with two axes of double refraction, and many of the laws of their phenomena, including the connection between optical structure and crystalline forms.
>4. The laws of metallic reflection.
>5. Experiments on the absorption of light.

He invented the kaleidoscope around 1815, the lenticular stereoscope, and the sea thermometer. He was chiefly responsible for the improvement of the British lighthouse system. Although Fresnel, who had also the satisfaction of being the first to put it into operation, perfected the dioptric apparatus independently, Brewster was active earlier in the field than Fresnel, describing the dioptric apparatus in 1812. He pressed its adoption on those in authority at least as early as 1820, two years before Fresnel suggested it. He was one of the leading contributors to the *Encyclopædia Britannica* (seventh and eighth editions) writing, among others, the articles on electricity, hydrodynamics, magnetism, microscope, optics, stereoscope, and voltaic electricity. He was author of the important biography, *Memoirs of the Life, Writings and Discoveries of Sir Isaac Newton* (1855). He was a man of highly honourable and fervently religious character. [source: Wikipedia bio] He was a devout Christian, and wrote "It cannot be presumption to be *sure* [of our forgiveness] because it is Christ's work, not ours; on the contrary, it is presumption to doubt his word and work." [source: Christian Influences in the Sciences, Daniel Graves]

**Friedrich Bessel** (1784-1846) Mathematician, astronomer, and systematizer of the Bessel functions (which were discovered by Daniel Bernoulli). Critical for the solution of certain differential equations, these functions are used throughout both classical and quantum physics. He worked on to produce precise positions for some 3,222 stars and published tables of atmospheric refraction: both based on James Bradley's stellar observations. The feat for which he is best remembered today is being the first to use parallax in calculating the distance to a star. [source: Wikipedia bio]

**William Buckland** (1784-1856) Geologist, palaeontologist and Dean of Westminster, who wrote the first full account of a fossil dinosaur, which he named Megalosaurus. He showed how detailed scientific analysis could be used to understand geohistory by reconstructing events from deep time. Buckland was a proponent of Old Earth creationism. In 1809, he was ordained as a priest (Anglican) and continued to make frequent geological excursions. In 1820 his *Vindiciæ Geologiæ; or the Connexion of Geology with Religion Explained* justified the new science of geology and reconciled geological evidence with the biblical accounts of creation and Noah's Flood. At a time when others were coming under the opposing influence of James Hutton's theory of uniformitarianism, Buckland developed a new hypothesis that the word "beginning" in Genesis meant an undefined period between the origin of the earth and the creation of its current inhabitants, during which a long series of extinctions and successive creations of new kinds of plants and animals had occurred. Thus, his catastrophism theory incorporated a version of Old Earth creationism or Gap creationism. Buckland initially believed in a global deluge during the time of Noah but was not a supporter of flood geology as he believed that only a small amount of the strata could have been formed in the single year occupied by the deluge. By 1840 he was very actively promoting the view that what had been interpreted as evidence of the "Universal Deluge" two decades earlier, and subsequently of deep submergence by a new generation of geologists such as Charles Lyell and Louis Agassiz, was in fact evidence of a major glaciation. Buckland shared the view of Georges Cuvier that no humans had coexisted with any extinct animals. He was commissioned to contribute one of the set of eight *Bridgewater Treatises*, "On the Power, Wisdom and Goodness of God, as manifested in the Creation". This took him almost five years' work and was published in 1836 with the title *Geology and Mineralogy considered with reference to Natural Theology*. His section was a detailed compendium of his theories of day-age, gap theory and theistic evolution. In the introduction he expressed the argument from design by asserting that the families and phyla of biology were "clusters of contrivance":

> The myriads of petrified Remains which are disclosed by the researches of Geology all tend to prove that our Planet has been occupied in times preceding the Creation of the Human Race, by extinct species of Animals and Vegetables, made up, like living Organic Bodies, of 'Clusters of Contrivances,' which demonstrate the exercise of stupendous Intelligence and Power. They further show that these extinct forms of Organic Life were so closely allied, by Unity in the principles of their construction, to Classes, Orders, and Families, which make up the existing Animal and Vegetable Kingdoms, that they not only afford an argument of surpassing force, against the doctrines of the Atheist and Polytheist; but

supply a chain of connected evidence, amounting to demonstration, of the continuous Being, and of many of the highest Attributes of the One Living and True God.

In response, computing pioneer Charles Babbage produced his *Ninth Bridgewater Treatise*. [source: Wikipedia bio]

**John James Audubon** (1785-1851) He was an ornithologist and naturalist who painted, catalogued, and described the birds of North America in a form far superior to what had gone before. His *Birds of America* consisted of 435 hand-colored, life-size prints of 497 bird species and described just over 700 North American bird species. Audubon's great work was a remarkable accomplishment. It took more than 14 years of field observations and drawings, plus his single-handed management and promotion of the project to make it a success. His field notes comprised a significant contribution to the understanding of bird anatomy and behavior. Among his accomplishments, Audubon discovered twenty-five new species and twelve new subspecies. In speaking of native Americans, he exhibited his belief in God: "Whenever I meet Indians, I feel the greatness of our Creator in all its splendor, for there I see the man naked from His hand and yet free from acquired sorrow." [source: Wikipedia bio]

**William Daniel Conybeare** (1787-1857) One of the most distinguished English geologists. He was ordained and became in 1814 curate of Wardington, near Banbury. Both Buckland and Sedgwick acknowledged their indebtedness to him for instruction received when they first began to devote attention to geology. In 1821, in collaboration with Henry De la Beche he distinguished himself by describing, from fragmentary remains, the saurian Plesiosaurus in a paper that also contained an important description and analysis of all that had been learned to that point about the anatomy of ichthyosaurs, including the fact that there had been at least three different species. His predictions about the plesiosaur were proved correct by the discovery of a nearly complete skeleton by Mary Anning in 1823. His principal work was the *Outlines of the Geology of England and Wales* (1822), being a second edition of the small work issued by William Phillips and written in co-operation with that author. The original contributions of Conybeare formed the principal portion of this edition. [source: Wikipedia bio]

**Augustin-Jean Fresnel** (1788-1827) He contributed significantly to the establishment of the theory of wave optics. Fresnel studied the behaviour of light both theoretically and experimentally. He is perhaps best known as the inventor of the Fresnel lens, first adopted in lighthouses. He was able to show via mathematical methods that polarization could be explained only if light was *entirely* transverse, with no longitudinal vibration whatsoever. [source: Wikipedia bio]

**Augustin-Louis Cauchy** (1789-1857) He was a mathematician and early pioneer of analysis. He started the project of formulating and proving the theorems of infinitesimal calculus in a rigorous manner. He also gave several important theorems in complex analysis and initiated the study of permutation groups in abstract algebra. His writings cover the entire range of mathematics and mathematical physics. He provided proof of Fermat's polygonal number theorem. In the theory of light he worked on Fresnel's wave theory and on the dispersion and polarization of light. He also contributed significant research in mechanics, substituting the notion of the continuity of geometrical displacements for the principle of the continuity of matter. He wrote on the equilibrium of rods and elastic membranes and on waves in elastic media. He introduced a $3 \times 3$ symmetric matrix of numbers that is now known as the Cauchy stress tensor. In elasticity, he originated the theory of stress, and his results are nearly as valuable as those of Simeon Poisson. Cauchy is most famous for his single-handed development of complex function theory. Also his well-known test for absolute convergence stems from his Cauchy condensation test. In 1829 he defined for the first time a complex function of a complex variable. He was a devout Catholic and lent his prestige and knowledge to the *École Normale Écclésiastique*, a school in Paris run by Jesuits, for training teachers for their colleges. He also took part in the founding of the *Institut Catholique*. The purpose of this institute was to counter the effects of the absence of Catholic university education in France. These activities did not make Cauchy popular with his colleagues who, on the whole, supported the Enlightenment ideals of the French Revolution. He was a member of the Society of Saint Vincent de Paul and also had links to the Society of Jesus and defended them at the Academy when it was politically unwise to do so. [source: Wikipedia bio]

**John Bachman** (1790-1874) American Lutheran minister, social activist and naturalist who collaborated with J. J. Audubon to produce *Viviparous Quadrapeds of North America* and whose writings, particularly *Unity of the Human Race*, were influential in the development of the theory of evolution. He served the same Charleston, South Carolina church as pastor for 56 years but still found time to conduct natural history studies that caught the attention of Audubon and eminent scientists in England, Europe, and beyond. Bachman was a social reformer who ministered to African-American slaves as well as white Southerners, and who used his knowledge of natural history to become one of the first writers to argue scientifically that blacks and whites are the same species. [source: Wikipedia bio]

**Michael Faraday** (1791-1867) Faraday studied the magnetic field around a conductor carrying a DC electric current, and established the basis for the electromagnetic field concept in physics. He discovered electromagnetic induction, diamagnetism, and laws of electrolysis. He established that magnetism could affect rays of light and that there was an underlying relationship between the two phenomena. His inventions of electromagnetic rotary devices formed the foundation of electric motor technology, and it was largely due to his efforts that electricity became viable for use in technology. As a chemist, Michael Faraday discovered benzene, invented an early form of the Bunsen burner and popularized terminology such as anode, cathode, electrode, and ion. He made the first rough experiments on the diffusion of gases. Faraday was the first to report what later came to be called metallic nanoparticles. He constructed an electric dynamo, the ancestor of modern power generators. He was one of the most influential scientists in history. Some historians of science refer to him as the best experimentalist in the history of science. Faraday was highly religious. Biographers have noted that "a strong sense of the unity of God and nature pervaded Faraday's life and work." [source: Wikipedia bio] For more on Faraday's deep religious piety, see: "The Christian Character of Michael Faraday as Revealed in His Personal Life and Recorded Sermons," by Phillip Eichman.

**Charles Babbage** (1791-1871) Mathematician, philosopher, inventor, and mechanical engineer who originated the concept of a programmable computer. Parts of his uncompleted mechanisms are on display in the London Science Museum. In 1991, a perfectly functioning difference engine was constructed from Babbage's original plans. The success of the finished engine indicated that Babbage's machine would have worked. Nine years later, the Science Museum completed the printer Babbage had designed for the difference engine, an astonishingly complex device for the 19th century. Considered a "father of the computer", Babbage is credited with inventing the first mechanical computer that eventually led to more complex designs. Babbage sought a method by which mathematical tables could be calculated mechanically, removing the high rate of human error. Three different factors seem to have influenced him: a dislike of untidiness; his experience working on logarithmic tables; and existing work on calculating machines. His machines were among the first mechanical computers, although they were not actually completed, largely because of funding problems and personality issues. He directed the building of some steam-powered machines that achieved some success, suggesting that calculations could be mechanised. Although Babbage's machines were mechanical and unwieldy, their basic architecture was very similar to a modern computer. He began in 1822 with what he called the difference engine, made to compute values of polynomial functions. Unlike similar efforts of the time, Babbage's difference engine was created to calculate a series of values automatically. By using the method of finite differences, it was possible to avoid the need for multiplication and division. He also started designing a different, more complex machine called the Analytical Engine that could be programmed using punched cards. In 1838, Babbage invented the pilot (also called a cow-catcher), the metal frame attached to the front of locomotives that clears the tracks of obstacles. He also invented an ophthalmoscope. In 1837, responding to the *Bridgewater Treatises*, of which there were eight, he published his *Ninth Bridgewater Treatise, "On the Power, Wisdom and Goodness of God, as manifested in the Creation"*, putting forward the thesis that God had the omnipotence and foresight to create as a divine legislator, making laws (or programs) which then produced species at the appropriate times, rather than continually interfering with *ad hoc* miracles each time a new species was required. The book is a work of natural theology, and incorporates extracts from correspondence he had been having with John Herschel on the subject. [source: Wikipedia bio]

**Sir John Herschel** (1792-1871) He named seven moons of Saturn and four moons of Uranus, investigated colour blindness and the chemical power of ultraviolet rays. Herschel made numerous important contributions to photography. He made improvements in photographic processes, particularly in inventing the cyanotype process and variations (such as the chrysotype), the precursors of the modern blueprint process. He experimented with color reproduction, noting that rays of different parts of the spectrum tended to impart their own color to a photographic paper. He discovered sodium thiosulfate to be a solvent of silver halides in 1819, and found that this "hyposulphite of soda" ("hypo") could be used as a photographic fixer, to "fix" pictures and make them permanent. [source: Wikipedia bio] He was won to sincere Christianity by the character of his wife. [source: *Christian Influences in the Sciences*, Daniel Graves]

**William Whewell** (1794-1866) He was a polymath, scientist, Anglican priest, philosopher, theologian, and historian of science.

What is most often remarked about Whewell is the breadth of his endeavours. At a time when men of science were becoming increasingly specialised, Whewell appears as a vestige of an earlier era when men of science dabbled in a bit of everything. He researched ocean tides (for which he won the Royal Medal), published work in the disciplines of mechanics, physics, geology, astronomy, and economics, while also finding the time to compose poetry, author a *Bridgewater Treatise*, translate the works of Goethe, and write sermons and theological tracts. His best-known works are two voluminous books which attempt to map and systematize the development of the sciences, *History of the Inductive Sciences* (1837) and *The Philosophy of the Inductive Sciences, Founded Upon Their History* (1840). Whewell came up with the term *scientist* itself (scientists had previously been known as "natural philosophers" or "men of science"). Whewell also contributed the terms *physicist*, *catastrophism*, and *uniformitarianism*, amongst others; Whewell suggested the terms *anode* and *cathode* to Michael Faraday. Whewell introduced what is now called the Whewell equation, an equation defining the shape of a curve without reference to an arbitrarily chosen coordinate system. [source: Wikipedia bio]

**Temple Chevallier** (1794-1873) Clergyman, astronomer, and mathematician. Between 1847 and 1849, he made important observations regarding sunspots. He made important observations of Jupiter's moons and regular meteorological observations. Not only did he write many papers on astronomy and physics, he also published a translation of the Apostolic Fathers that went into a second edition, and translated the works of Clement of Alexandria, Polycarp and Ignatius of Antioch. His lectures were published in 1835 as *Of the proofs of the divine power and wisdom derived from the study of astronomy*. [source: Wikipedia bio]

**Johann Heinrich von Mädler** (1794-1874) In 1830, he and Joseph von Fraunhofer began producing drawings of Mars: the first true maps of that planet. They made a preliminary determination for Mars' rotation period, which was off by almost

13 seconds. A later determination in 1837 was off by only 1.1 seconds. They also produced the first exact map of the Moon, *Mappa Selenographica*, published in four volumes in 1834–1836. In 1837 a description of the Moon (*Der Mond*) was published. Both were the best descriptions of the Moon for many decades, not superseded until the map of Johann Friedrich Julius Schmidt in the 1870s. He determined that the features on the Moon do not change, and that it had no atmosphere or water. [source: Wikipedia bio]

**John Stevens Henslow** (1796-1861) Botanist and geologist, who made valuable observations on the geology of the Isle of Man and parts of Anglesey. Henslow had studied mineralogy with considerable zeal, and in 1822 was appointed Professor of Mineralogy in the University of Cambridge. Two years later he took holy orders. In 1843 he discovered nodules of coprolitic origin in the Red Crag at Felixstowe in Suffolk, and two years later he called attention to those also in the Cambridge Greensand and remarked that they might be of use in agriculture. Although Henslow derived no benefit, these discoveries led to the establishment of the phosphate industry in Suffolk and Cambridgeshire; and the works proved lucrative until the introduction of foreign phosphates. [source: Wikipedia bio]

**Thomas Hodgkin** (1798-1866) He was one of the most prominent pathologists of his time and a pioneer in preventive medicine. He is now best known for the first account of Hodgkin's disease, a form of lymphoma and blood disease, in 1832. His 1829 two-volume work, *The Morbid Anatomy of Serous and Mucous Membranes*, became a classic in modern pathology. Hodgkin was one of the earliest defenders of preventive medicine, having published *On the Means of Promoting and Preserving Health* in book form in 1841. Among other early observations were the first description of acute appendicitis, of the biconcave format of red blood cells and the striation of muscle fibers. He was a Quaker. [source: Wikipedia bio]

**Hugh Miller** (1802-1856) He was a self-taught geologist and evangelical Christian. Among his geological works are *Footprints of the Creator* (1850) and *The Testimony of the Rocks; or, Geology in its Bearings on the Two Theologies, Natural and Revealed* (1856). Miller held that the Earth was of great age, and that it had been inhabited by many species which had come into being and gone extinct, and that these species were homologous; although that species were progressing with time. He denied the Epicurean theory that new species occasionally budded from the soil, and the Lamarckian theory of development of species, as lacking evidence. He argued that all this showed the direct action of a benevolent Creator, as attested in the Bible - the similarities of species are manifestations of types in the Divine Mind; he accepted the view of Thomas Chalmers that Genesis begins with an account of geological periods, and does not mean that each of them is a day; Noah's Flood was a limited subsidence of the Middle East. Geology, to Miller, offers a better version of the argument from design than William Paley could provide, and answers the objections of skeptics, by showing that living species did not arise by chance or by impersonal law. Though he had no academic credentials, he is today considered one of Scotland's premier paleontologists. [source: Wikipedia bio]

**Christian Andreas Doppler** (1803-1853) He is most famous for what is now called the Doppler effect, which is the apparent change in frequency and wavelength of a wave as perceived by an observer moving relative to the wave's source. Doppler, along with Franz Unger, played an influential role in the development of young Gregor Mendel, known as the founding father of genetics, who was a student at the University of Vienna from 1851 to 1853. [source: Wikipedia bio]

**Louis Agassiz** (1807-1873) He was a paleontologist, glaciologist, geologist, ichthyologist, and a prominent innovator in the study of the Earth's natural history. The work that laid the foundation of his worldwide fame, was his magnificently illustrated *Recherches* *sur les poissons fossiles* (*Research on Fossil Fish*), published in five volumes from 1833 to 1843. In 1837 Agassiz was the first to scientifically propose that the Earth had been subject to a past ice age. His 1840 two-volume work, *Etudes sur les glaciers* (*Study on Glaciers*) discussed the movements of the glaciers, their moraines, and their influence in grooving and rounding the rocks over which they traveled. It gave a fresh impetus to the study of glacial phenomena in all parts of the world. Agassiz remarked "that great sheets of ice, resembling those now existing in Greenland, once covered all the countries in which unstratified gravel (boulder drift) is found; that this gravel was in general produced by the trituration of the sheets of ice upon the subjacent surface, etc." In 1846 he delivered in America twelve lectures on "The Plan of Creation as shown in the Animal Kingdom." Agassiz is also remembered today for his resistance to Charles Darwin's theories on evolution, which lasted his entire life. [source: Wikipedia bio]

## Chapter Seven

## 56 Catholic, Protestant and Otherwise Religious Prominent Scientists: 1850-1900 (From Maxwell to Mendel, Pasteur, and Kelvin)

**Adam Sedgwick** (1785-1873) One of the founders of modern geology. He proposed the Devonian period of the geological timescale and later the Cambrian period. Though he had guided the young Charles Darwin in his early study of geology, Sedgwick was an outspoken opponent of Darwin's theory of evolution by means of natural selection. He investigated the phenomena of metamorphism and concretion, and was the first to distinguish clearly between stratification, jointing, and slaty cleavage. His geological position was catastrophist in the mid 1820s, but following Charles Lyell's 1830 publication of uniformitarian ideas he came to accept that a worldwide flood was untenable and talked of floods at various dates. He strongly believed that species of organisms originated in a succession of Divine creative acts throughout the long expanse of history. While he became increasingly Evangelical with age, he strongly supported advances in geology. When Robert Chambers anonymously published his own theory of universal evolutionism as his "development hypothesis" in the book *Vestiges of the Natural History of Creation* published in October 1844 to immediate popular success, Sedgwick's many friends urged him to respond. Like other eminent scientists he initially ignored the book, but the subject kept recurring and he then read it carefully and made a withering attack on the book in the July 1845 edition of the *Edinburgh Review*. Sedgwick never accepted the case for evolution made in *On the Origin of Species* in 1859. In response to receiving and reading Darwin's book, he wrote to Darwin saying:

> If I did not think you a good tempered & truth loving man I should not tell you that. . . . I have read your book with more pain than pleasure. Parts of it I admired greatly; parts I laughed at till my sides were almost sore; other parts I read with absolute sorrow; because I think them utterly false & grievously mischievous — You have deserted—after a start in that tram-road of all solid physical truth—the true method of induction—& started up a machinery as wild I think as Bishop Wilkin's locomotive that was to sail with us to the Moon. Many of your wide conclusions are based upon assumptions which can neither be proved nor disproved. Why then express them in the language & arrangements of philosophical induction?

Despite this difference of opinion, the two men remained friendly until Sedgwick's death. His opposition seems linked, not to his religion as such, but to the particular cast of his beliefs. [source: Wikipedia bio]

**Edward Hitchcock** (1793-1864) His chief project was natural theology, which attempted to unify and reconcile science and religion, focusing on geology. His major work in this area was *The Religion of Geology and its Connected Sciences* (Boston, 1851). He knew that the earth was at least hundreds of thousands of years old. He explicitly rejected not only atheistic evolution, but a six day creation as well. He thought that new species were introduced by a Deity at the right time in the history of the earth. In 1863 Hitchcock wrote an article highly critical of Darwin's theory of natural selection. [source: Wikipedia bio]

Sir Charles Lyell (1797-1875) He was the foremost geologist of his day and is best known as the author of *Principles of Geology*, which popularised uniformitarianism – the idea that the earth was shaped by slow-moving forces still in operation today. The central argument in *Principles* was that *the present is the key to the past*. Geological remains from the distant past can, and should, be explained by reference to geological processes now in operation and thus directly observable. Lyell's interpretation of geologic change as the steady accumulation of minute changes over enormously long spans of time was a powerful influence on the young Charles Darwin. They became good friends. His geological interests ranged from geological dynamics through stratigraphy, paleontology, and glaciology to topics that would now be classified as prehistoric archaeology and paleoanthropology. One of the contributions that Lyell made in *Principles* was to explain the cause of earthquakes; he made observations about surface irregularities such as faults, fissures, stratigraphic displacements and depressions. In his work on volcanoes he supported gradual building of volcanoes, so-called "backed up-building". Lyell's most important specific work was in the field of stratigraphy. He concluded that recent strata (rock layers) could be categorized according to the number and proportion of marine shells encased within. Based on this he proposed dividing the Tertiary period into three parts, which he named the Pliocene, Miocene, and Eocene. His observational methods and general analytical framework remain in use today as foundational principles in geology. Lyell, a devout Christian, had great difficulty, however, reconciling his beliefs with natural selection. He had proposed "Centres of Creation" to explain diversity and territory of species, but he was fairly open to the idea of evolution. In his *Geological Evidences of the Antiquity of Man* (1863; four years after Darwin's *Origin of Species*) his acceptance of natural selection and evolution was equivocal at best. It acknowledged that human bodies might have evolved, but left open the possibility of divine intervention in the origins of human intellect and moral sense. When Lyell wrote that it remained a profound mystery how the huge gulf between man and beast could be bridged, Darwin wrote "Oh!" in the margin of his copy. Darwin wrote to his associate Huxley about the book: "I am fearfully disappointed at Lyell's excessive caution." [source: Wikipedia bio]

**Joseph Henry** (1797-1878) He discovered the electromagnetic phenomenon of self-inductance and also mutual inductance (independently of Michael Faraday). Henry's work on the electromagnetic relay was the basis of the electrical telegraph. He was the first to coil insulated wire tightly around an iron core in order to make a more powerful electromagnet. He also showed that, when making an electromagnet using just two electrodes attached to a battery, it is best to wind several coils of wire in parallel, but when using a set-up with multiple batteries, there should be only one single long coil. The latter made the telegraph feasible. In 1831 he created one of the first machines to use electromagnetism for motion. This was the earliest ancestor of modern DC motor. Henry identified the room acoustics phenomena we now call direct sound, early reflections, and reverberation. He demonstrated the early sound integration period and laid the groundwork for further fundamental research on early reflections that was not followed up until the work at Göttingen University in the 1950–1960s. He brought a robust scientific approach to the subject of acoustics. [source: Wikipedia bio]

**Sir George Biddell Airy** (1801-1892) Astronomer Royal from 1835 to 1881. Airy's discovery of a new inequality in the motions of Venus and the Earth is in some respects his most remarkable achievement. The investigation was probably the most laborious that had been made up to Airy's time in planetary theory, and represented the first specific improvement in the solar tables effected in England since the establishment of the theory of gravitation. Another was his determination of the mean density of the Earth. In 1826, the idea occurred to him of attacking this problem by means of pendulum experiments at the top and bottom of a deep mine. The experiments eventually took place at the Harton pit near South Shields in 1854. Their immediate result was to show that gravity at the bottom of the mine exceeded that at the top by 1/19286 of its amount, the depth being 383 m (1,256 ft) From this he was led to the final value of Earth's specific density of 6.566. The currently accepted value for Earth's density is 5.5153 g/cm³. In 1862, Airy presented a new technique to determine the strain and stress field within a beam. This technique, sometimes called the Airy stress function method, can be used to find solutions to many two-dimensional problems in solid mechanics. [source: Wikipedia bio]

Sir Richard Owen (1804-1892) Biologist, comparative anatomist and palaeontologist. He was a pioneer in concise anatomical nomenclature and, so far at least as the vertebrate skeleton is concerned, he first clearly distinguished between the now-familiar phenomena of analogy and homology. He made important contributions to every department of comparative anatomy and zoology, for a period of over fifty years. Owen is probably best remembered today for coining the word *Dinosauria* (meaning "Terrible Reptile") and for his outspoken opposition to Charles Darwin's theory of evolution by natural selection. He agreed with Darwin that evolution occurred, but thought it was more complex than outlined in Darwin's *Origin*. He was the driving force behind the establishment, in 1881, of the British Museum (Natural History) in London. After the publication of Charles Darwin's *On The Origin of Species*, Owen had grave doubts that transmutation would bestialize man. In April 1860 the *Edinburgh Review* published Owen's anonymous review of the *Origin*. In it Owen objected to what he saw as Darwin's caricature of the creationist position, and his ignoring Owen's "axiom of the continuous operation of the ordained becoming of living things". He thought that the book symbolized a sort of "abuse of science." The two clashed over the next decade to such an extent that Darwin exclaimed in 1871: "I used to be ashamed of hating him so much, but now I will carefully cherish my hatred and contempt to the last days of my life". [source: Wikipedia bio]

Thomas Graham (1805-1869) He is best known for two things: 1) His studies on the diffusion of gases resulted in "Graham's Law", which states that the rate of effusion of a gas is inversely proportional to the square root of its molar mass. 2) His discovery of dialysis, which is used in many medical facilities today, was the result of Graham's study of colloids. This work resulted in Graham's ability to separate colloids and crystalloids using a so-called "dialyzer", the precursor of today's dialysis machine. This study initiated the scientific field known as colloid chemistry, of which Graham is credited as the founder. [source: Wikipedia bio]

**Johann von Lamont** (1805-1879) He created a star catalog that had about 35,000 entries. He discovered a magnetic decennial period (ten-year cycle) and the electric current in the Earth closing the electric "circuit" creating the magnetic field in 1850. This roughly matched the eleven-year sunspot cycle discovered by Heinrich Schwabe. He calculated the orbits of the moons of Uranus and Saturn, obtaining the first value for Uranus' mass. [source: Wikipedia bio]

**Matthew Maury** (1806-1873) Astronomer, oceanographer, meteorologist, cartographer, and geologist. He was nicknamed Pathfinder of the Seas and Father of modern Oceanography and Naval Meteorology and later, Scientist of the Seas, due to the publication of his extensive works in his books, especially *Physical Geography of the Sea* (1855): the first extensive and comprehensive book on oceanography to be published. Maury made many important new contributions to charting winds and ocean currents, including ocean lanes for passing ships at sea. He published the Wind and Current Chart of the North Atlantic, which showed sailors how to use the ocean's currents and winds to their advantage and drastically reduced the length of ocean voyages. Maury's uniform system of recording oceanographic data was adopted by navies and merchant marines around the world and was used to develop charts for all the major trade routes. He was convinced that adequate scientific knowledge of the sea could be obtained only through international cooperation. He proposed that the United States invite the maritime nations of the world to a conference to establish a "universal system" of meteorology, and he was the leading spirit of that pioneer scientific conference when it met in Brussels in 1853. Within a few years, nations owning three fourths of the shipping of the world were sending their oceanographic observations to Maury at the Naval Observatory, where the information was evaluated and the results given worldwide distribution. He later gave talks in Europe about cooperation on a weather bureau for land just as he had charted the winds and predicted storms at sea many years before. Maury lived by the Scriptures; he fully and unconditionally believed in what the Holy Scriptures stated; he hardly ever spoke or wrote without the inclusion of scriptural references; he prayed every day. [source: Wikipedia bio] He created the science of oceanography because he believed the Bible when it said there were paths in the seas. His Christian belief ran deep and touched all he did. [source: *Christian Influences in the Sciences*, Daniel Graves]

**Arnold Henry Guyot** (1807-1884) He ranked high as a geologist and meteorologist. As early as 1838, he undertook, at Agassiz's suggestion, the study of glaciers, and was the first to announce, in a paper submitted to the Geological Society of France, certain important observations relating to glacial motion and structure. Among other things he noted the more rapid flow of the center than of the sides, and the more rapid flow of the top than of the bottom of glaciers; described the laminated or ribboned structure of the glacial ice, and ascribed the movement of glaciers to a gradual molecular displacement rather than to a sliding of the ice mass. His extensive meteorological observations in America led to the establishment of the United States Weather Bureau, and his *Meteorological and Physical Tables* (1852, revised ed. 1884) were long standard. [source: Wikipedia bio]

**Charles Darwin** (1809-1882) He was the originator of the theory of evolution based on natural selection. Later in life he became an agnostic, but he never seems to have outright denied God's existence or to have claimed atheism as his position. At the time his famous theory was first published in 1859, he was undeniably a theist: thus demonstrating yet again that a theistic -- not an atheistic -- mind came up with an important new scientific understanding or discovery. Originally he had planned to become an Anglican clergyman. Here is the evidence for his theism (of some sort; in a sub-Christian sense) until very late in life:

> With respect to the theological view of the question; this is always painful to me.— I am bewildered.— I had no intention to write atheistically. But I own that I cannot see, as plainly as others do, & as I should wish to do, evidence of design & beneficence on all sides of us. There seems to me too much misery in the world. I cannot persuade myself that a beneficent & omnipotent God would have designedly created the Ichneumonidæ with the express intention of their feeding within the living bodies of caterpillars, or that a cat should play with mice. Not believing this, I see no necessity in the belief that the eye was expressly designed. On the other hand I cannot anyhow be contented to view this wonderful universe & especially the nature of man, & to conclude that everything is the result of brute force. I am inclined to look at everything as resulting from designed laws, with the details, whether good or bad, left to the working out of what we may call chance. Not that this notion *at all* satisfies me. I feel most deeply that the whole subject is too profound for the human intellect. A dog might as well speculate on the mind of Newton.— Let each man hope & believe what he can.—
>
> Certainly I agree with you that my views are not at all necessarily atheistical. . . . I can see no reason, why a man, or other animal, may not have been aboriginally produced by other laws; & that all these laws may have been expressly designed by an omniscient Creator, who foresaw every future event & consequence. But the more I think the more bewildered I become; as indeed I have probably shown by this letter.
>
> (Letter to Christian botanist Asa Gray, 22 May 1860)
>
> I am myself quite conscious that my mind is in simple muddle about "designed laws" & "undesigned

consequences". Does not Kant say that there are several subjects on which directly opposite conclusions can be proved true?! . . . Yet, as I said before, I cannot persuade myself that electricity acts, that the tree grows, that man aspires to loftiest conceptions all from blind, brute force.

(Letter to Asa Gray, 3 July 1860)

With respect to Design, I feel more inclined to show a white flag than to fire my usual long-range shot. . . . If anything is designed, certainly Man must be; one's "inner consciousness" (though a false guide) tells one so; . . . You say that you are in a haze; I am in thick mud;—the orthodox would say in fetid abominable mud. I believe I am in much the same frame of mind as an old Gorilla would be in if set to learn the first book of Euclid. The old Gorilla would say it was of no manner of use; & I am much of the same mind; yet I cannot keep out of the question.

(Letter to Asa Gray, 11 December 1861)

I may say that the impossibility of conceiving that this grand and wondrous universe, with our conscious selves, arose through chance, seems to me the chief argument for the existence of God; but whether this is an argument of real value, I have never been able to decide. . . . The safest conclusion seems to be that the whole subject is beyond the scope of man's intellect . . .

(Letter to N. D. Doedes, 2 April 1873)

Another source of conviction in the existence of God, connected with the reason, and not with the feelings, impresses me as having much more weight. This follows from the extreme difficulty or rather impossibility of conceiving this immense and wonderful universe, including man with his capacity of looking far backwards and far into futurity, as the result of blind chance or necessity. When thus reflecting I feel compelled to look to a First Cause having an intelligent mind in some degree analogous to that of man; and I deserve to be called a Theist. This conclusion was strong in my mind about the time, as far as I can remember, when I wrote the 'Origin of Species;' and it is since that time that it has very gradually, with many fluctuations, become weaker. . . .

I cannot pretend to throw the least light on such abstruse problems. The mystery of the beginning of all things is insoluble by us; and I for one must be content to remain an Agnostic.

(*Autobiography*, 1876)

I most wholly agree with you that there is no reason why the disciples of either school should attack each other with bitterness, though each upholding strictly their beliefs.

You, I am sure, have always practically acted in this manner in your conduct towards me & I do not doubt to all others. Nor can I remember that I have ever published a word directly against religion or the clergy.

(Letter to John Brodie Innes, 27 November 1878)

It seems to me absurd to doubt that a man may be an ardent Theist & an evolutionist.— You are right about Kingsley. Asa Gray, the eminent botanist, is another case in point— What my own views may be is a question of no consequence to any one except myself.— But as you ask, I may state that my judgment often fluctuates. Moreover whether a man deserves to be called a theist depends on the definition of the term: which is much too large a subject for a note. In my most extreme fluctuations I have never been an atheist in the sense of denying the existence of a God.— I think that generally (& more and more so as

I grow older) but not always, that an agnostic would be the most correct description of my state of mind.
(Letter to John Fordyce, 7 May 1879 [complete] )

Nevertheless you have expressed my inward conviction, though far more vividly and clearly than I could have done, that the Universe is not the result of chance. But then with me the horrid doubt always arises whether the convictions of man's mind, which has been developed from the mind of the lower animals, are of any value or at all trustworthy. Would any one trust in the convictions of a monkey's mind, if there are any convictions in such a mind?

(Letter to William Graham, 3 July 1881)

[for more on this topic, see: John Hedley Brooke, "Charles Darwin on Religion"; "What Did Darwin Believe?"; "Darwin and the Church"]

**Asa Gray** (1810-1888) He is considered the most important American botanist of the 19th century. A devout Christian, and considered by Darwin to be his friend and "best advocate", he attempted to convince Darwin that design was inherent in all forms of life, and to return to his faith. Darwin agreed that his theories were "not at all necessarily atheistical" but was unable to share Gray's belief. Gray was a staunch supporter of Darwin in America, and collected together a number of his own writings to produce an influential book, *Darwiniana*. These essays argued for a conciliation between Darwinian evolution and the tenets of theism, at a time when many on both sides perceived the two as mutually exclusive. Gray denied that investigation of physical causes stood opposed to the theological view and the study of the harmonies between mind and Nature, and thought it "most presumable that an intellectual conception realized in Nature would be realized through natural agencies." [source: Wikipedia bio]

**Philip Henry Gosse** (1810-1888) Naturalist and popularizer of natural science, virtually the inventor of the seawater aquarium, and a painstaking innovator in the study of marine biology. Gosse is perhaps best known today as the author of *Omphalos*: an attempt to reconcile the immense geological ages presupposed by Charles Lyell with the biblical account of creation. In 1832 Gosse experienced a religious conversion—as he said, "solemnly, deliberately and uprightly, took God for my God." In 1860 he published *Actinologia Britannica*. Reviewers especially praised the color lithographs made from Gosse's watercolors. The *Literary Gazette* said that Gosse now stood "alone and unrivalled in the extremely difficult art of drawing objects of zoology so as to satisfy the requirements of science" as well as providing "vivid aesthetic impressions." [source: Wikipedia bio]

**Sir James Young Simpson** (1811-1870) He discovered the anaesthetic properties of chloroform and successfully introduced it for general medical use. In 1847 Simpson discovered the properties of chloroform during an experiment with friends in which he learnt that it could be used to put one to sleep. He improved the design of obstetric forceps and introduced anaesthesia to childbirth. He was a very early advocate of the use of midwives in the hospital environment. Many prominent women also consulted him for their gynaecological problems. [source: Wikipedia bio] Simpson was an ardent New Presbyterian. Asked by a reporter what was his greatest discovery, he replied, "When I learned Jesus Christ had died for my sins." [source: *Christian Influences in the Sciences*, Daniel Graves]

**Urbain Le Verrier** (1811-1877) He specialized in celestial mechanics and is best known for his part in the discovery of Neptune, using only mathematics and astronomical observations of the known planet Uranus. In 1859, Le Verrier was the first to report that the slow precession of Mercury's orbit around the Sun could not be completely explained by Newtonian mechanics and perturbations by the known planets. The anomalous precession was eventually explained by general relativity theory. He was a practicing Catholic. [source: Wikipedia bio]

**James Dwight Dana** (1813-1895) Geologist, mineralogist and zoologist, who made important studies of mountain-building, volcanic activity, and the origin and structure of continents and oceans. He was the pre-eminent American geologist of his time. Dana was responsible for developing much of the early knowledge on Hawaiian volcanism. In 1880 and 1881 he led the first geological study of the volcanics of Hawaii island. Dana's best known books were his *System of Mineralogy* (1837), his *Manual of Mineralogy* (1848), and his *Manual of Geology* (1863). A bibliographical list of his writings shows 214 titles of books and papers. He published a number of manuscripts in an effort to reconcile scientific findings with the Bible between 1856 and 1857 and which are called *Science and the Bible*. [source: Wikipedia bio] He was converted in a revival and lived an impeccable life thereafter. [source: *Christian Influences in the Sciences*, Daniel Graves]

**George Boole** (1815-1864) Mathematician and philosopher. As the inventor of Boolean logic—the basis of modern digital computer logic—Boole is regarded in hindsight as a founder of the field of computer science. His well-known *Treatise on Differential Equations* appeared in 1859, and was followed, the next year, by a *Treatise on the Calculus of Finite Differences*, designed to serve as a sequel to the former work. These treatises are valuable contributions to the important branches of mathematics in question. His reflections upon scientific, philosophical and religious questions are contained in four addresses upon *The Genius of Sir Isaac Newton*, *The Right Use of Leisure*, *The Claims of Science* and *The Social Aspect of Intellectual Culture*. Claude Shannon recognised that Boole's work could form the basis of mechanisms and processes in the real world and in 1937 wrote his master's thesis at the Massachusetts Institute of Technology, in which he showed how Boolean algebra could optimize the design of systems of electromechanical relays, then used in telephone routing switches. He also proved that circuits with relays could solve Boolean algebra problems. Employing the properties of electrical switches to process logic is the basic concept that underlies all modern electronic digital computers. Victor Shestakov at Moscow State University (1907–1987) proposed a theory of electric switches based on Boolean logic even earlier, in 1935. Hence Boolean algebra became the foundation of practical digital circuit design; and Boole, via Shannon and Shestakov, provided the theoretical grounding for the Digital Age. He held and practiced the Christian faith. [source: Wikipedia bio]

**Crawford Williamson Long** (1815-1878) Physician and pharmacist best known for being the first to use diethyl ether as an anesthetic in amputations and childbirth, and various operations. [source: Wikipedia bio]

**Karl Weierstrass** (1815-1897) He was a mathematician and is often cited as the "father of modern analysis". Weierstrass was interested in the soundness of calculus. At the time, there were somewhat ambiguous definitions regarding the foundations of calculus, and hence important theorems could not be proven with sufficient rigour. Using the concept of uniform convergence, Weierstrass was able to write proofs of several then-unproven theorems such as the intermediate value theorem, the Bolzano-Weierstrass theorem, and the Heine-Borel theorem. He also made significant advancements in the field of calculus of variations. Using the apparatus of analysis that he helped to develop, Weierstrass was able to give a complete reformulation of the theory which gave way for the modern study of calculus of variations. He was a Catholic. [source: Wikipedia bio]

**Sir Joseph Henry Gilbert** (1817-1901) He was noteworthy for his long career spent improving the methods of practical agriculture. The work which he carried out for over fifty years, in collaboration with John Bennet Lawes was of a most comprehensive character, involving the application of many branches of science, such as chemistry, meteorology, botany, animal and vegetable physiology, and geology; and its influence in improving the methods of practical agriculture extended all over the civilized world. [source: Wikipedia bio]

**Pietro Angelo Secchi** (1818-1878; Jesuit priest) He was a pioneer in astronomical spectroscopy, and was one of the first scientists to state authoritatively that the Sun is a star. He drew some of the first color illustrations of Mars and was the first to refer to *canali*, the Italian word for *channels*, on the planetary surface. He proved that the solar corona and coronal prominences observed during a solar eclipse were part of the Sun, and not artifacts of the eclipse. He discovered solar spicules. He invented the heliospectrograph, star spectrograph, and telespectroscope. He showed that certain absorption lines in the spectrum of the Sun were caused by absorption in the Earth's atmosphere. He developed the first system of stellar classification: the five Secchi classes. He invented the Secchi disk, which is used to measure water transparency in oceans and lakes. [source: Wikipedia bio]

**James Prescott Joule** (1818-1889) He studied the nature of heat, and discovered its relationship to mechanical work (see energy). This led to the theory of conservation of energy, which led to the development of the first law of thermodynamics. He made observations on magnetostriction, and found the relationship between the current through a resistance and the heat dissipated, now called Joule's law. He once wrote: "Believing that the power to destroy belongs to the Creator alone I affirm . . . that any theory which, when carried out, demands the annihilation of force, is necessarily erroneous." He collaborated with Lord Kelvin from 1852 to 1856, and the resultant discoveries, including the Joule-Thomson effect, and the published results did much to bring about general acceptance of Joule's work and the kinetic theory. His gravestone is inscribed with the number "772.55", his climacteric 1878 measurement of the mechanical equivalent of heat, and with a quotation from the Gospel of John, "I must work the works of him that sent me, while it is day: the night cometh, when no man can work" (9:4). [source: Wikipedia bio]

**Léon Foucault** (1819-1868) Best known for the invention of the Foucault pendulum, a device demonstrating the effect of the Earth's rotation. He also made an early measurement of the speed of light and discovered eddy currents. In 1850, he did an experiment using the Fizeau–Foucault apparatus to measure the speed of light; it was viewed as "driving the last nail in the coffin" of Newton's corpuscle theory of light when it showed that light travels more slowly through water than through air. In 1851, he provided the first experimental demonstration of the rotation of the Earth on its axis. This was achieved by considering the rotation of the plane of oscillation of a freely suspended, long and heavy pendulum. [source: Wikipedia bio]

**Thomas Anderson** (1819-1874) Discovered the correct structure for codeine. In 1868 he discovered pyridine and related organic compounds such as picoline. As well as his work on organic chemistry, Anderson made important contributions to agricultural chemistry, writing over 130 reports on soils, fertilisers and plant diseases. He kept abreast of all areas of science, and was able to advise his colleague Joseph Lister on Pasteur's germ theory and the use of carbolic acid as an antiseptic. [source: Wikipedia bio]

**John Couch Adams** (1819-1892) Mathematician and astronomer. His most famous achievement was predicting the existence and position of Neptune, using only mathematics. The calculations were made to explain discrepancies with Uranus's orbit and the laws of Kepler and Newton. At the same time, but unknown to each other, the same calculations were made by Urbain Le Verrier. He also did much important work on gravitational astronomy and terrestrial magnetism. He was particularly adept at fine numerical calculations, often making substantial revisions to the contributions of his predecessors. In 1852, he published new and accurate tables of the moon's parallax. Adams ascertained that the Leonid cluster of meteors, which belongs to the solar system, traverses an elongated ellipse in 33.25 years, and is subject to definite perturbations from the larger planets, Jupiter, Saturn and Uranus. A Wesleyan, he won college prizes for Bible studies. [source: Wikipedia bio]

**Armand Hippolyte Louis Fizeau** (1819-1896) In 1848, he predicted the redshifting of electromagnetic waves. In 1849 he published the first results obtained by his method for determining the speed of light (see Fizeau-Foucault apparatus), and in 1850 with E. Gounelle measured the speed of electricity. Hippolyte in 1864 made the first suggestion that the "length of a light wave be used as a length standard". In 1853 he described the use of the capacitor (then called the *condenser*) as a means to increase the efficiency of the induction coil. [source: Wikipedia bio]

**Charles Piazzi Smyth** (1819-1900) Astronomer Royal for Scotland from 1846 to 1888, well-known for many innovations in astronomy. He was the pioneer of the modern practice of placing telescopes at high altitudes to enjoy the best observing conditions. Smyth investigated the spectra of the aurora, and zodiacal light. He recommended the use of the rain-band for weather forecasting and discovered, in conjunction with Alexander Stewart Herschel, the harmonic relation between the rays emitted by carbon monoxide. [source: Wikipedia bio]

**George Gabriel Stokes** (1819-1903) Studied the steady motion of incompressible fluids and some cases of fluid motion, the friction of fluids in motion, the equilibrium and motion of elastic solids, and the internal friction of fluids on the motion of pendulums. To the theory of sound he made several contributions, including a discussion of the effect of wind on the intensity of sound and an explanation of how the intensity is influenced by the nature of the gas in which the sound is produced. These inquiries together put the science of fluid dynamics on a new footing, and provided a key not only to the explanation of many natural phenomena, such as the suspension of clouds in air, and the subsidence of ripples and waves in water, but also to the solution of practical problems, such as the flow of water in rivers and channels, and the skin resistance of ships. His work on fluid motion and viscosity led to his calculating the terminal velocity for a sphere falling in a viscous medium. This became known as Stokes' law. He derived an expression for the frictional force (also called drag force) exerted on spherical objects with very small Reynolds numbers. His work is the basis of the falling sphere viscometer, in which the fluid is stationary in a vertical glass tube. Perhaps his best-known researches are those that deal with the wave theory of light. His first papers on the aberration of light appeared in 1845 and 1846, and were followed in 1848 by one on the theory of certain bands seen in the spectrum. In 1849 he published a long paper on the dynamical theory of diffraction, in which he showed that the plane of polarization must be perpendicular to the direction of propagation. In his famous paper on the change of wavelength of light, he described the phenomenon of fluorescence. He enunciated the fundamental principles of spectroscopy. He did much to advance mathematical physics (including Stokes' theorem). Stokes held conservative religious values and beliefs and published a volume on natural theology. He was vice president of the British and Foreign Bible Society and was active in foreign missions doctrinal issues. [source: Wikipedia bio] He rejected Darwin's theory of evolution, saying it was based on inadequate evidence. He had learned to read by reading the Psalms. As a Christian he said that evidence for Christ's resurrection must lead to action commensurate with the fact. [source: *Christian Influences in the Sciences*, Daniel Graves]

**Sir John William Dawson** (1820-1899) He entered zealously into studying the geology of Canada. With Charles Lyell in 1852, he obtained the first remains of an air-breathing reptile named Dendrerpeton. He also described the fossil plants of the Silurian, Devonian and Carboniferous rocks of Canada. In his books on geological subjects he maintained a distinctly theological attitude, declining to admit the descent or evolution of man from brute ancestors, and holding that the human species only made its appearance on this earth within quite recent times. [source: Wikipedia bio]

**Rudolf Virchow** (1821-1902) Anthropologist, pathologist, and biologist, known for his advancement of public health. Referred to as "the father of modern pathology," he is considered one of the founders of social medicine. Virchow's most widely known scientific contribution is his cell theory, which built on the work of Theodor Schwann. He is cited as the first to recognize leukemia cells. Virchow is also famous for elucidating the mechanism of pulmonary thromboembolism, coining the term *embolism*. Related to this research, Virchow described the factors contributing to venous thrombosis, Virchow's triad. Furthermore, Virchow founded the medical fields of cellular pathology and comparative pathology (comparison of diseases common to humans and animals). He also developed a standard method of autopsy procedure, named for him, that is still one of the two main techniques used today. Virchow famously delivered an anti-Darwinian lecture on human and primate skulls, in which he emphasized the lack of fossil evidence for a common ancestor of man and ape. [source: Wikipedia bio]

**Gregor Johann Mendel** (1822-1884; priest) He gained posthumous fame as the figurehead of the new science of genetics for his study of the inheritance of certain traits in pea plants. Mendel showed that the inheritance of these traits follows particular laws. He also studied astronomy and meteorology, founding the "Austrian Meteorological Society" in 1865. The majority of his published works were related to meteorology. Between 1856 and 1863 Mendel cultivated and tested some 29,000 pea plants (*i.e., Pisum sativum*). This study showed that one in four pea plants had purebred recessive alleles, two out of four were hybrid and one out of four were purebred dominant. His experiments led him to make two generalizations, the Law of Segregation and the Law of Independent Assortment, which later became known as Mendel's Laws of Inheritance. Mendel did read his paper, *Experiments on Plant Hybridization*, at two meetings of the Natural History Society of Brünn in Moravia in 1865. When Mendel's paper was published in 1866 in *Proceedings of the Natural History Society of Brünn*, it had little impact and was cited about three times over the next thirty-five years. His paper was criticized at the time, but is now considered a seminal work. At first Mendel's work was rejected, and it was not widely accepted until after he died. His ideas were rediscovered in the early twentieth century, and in the 1930s and

1940s the modern synthesis combined Mendelian genetics with Darwin's theory of natural selection. [source: Wikipedia bio] Mendel did not accept Darwin's theory, because his own discoveries in genetics showed that creatures tend to revert to kind. [source: *Christian Influences in the Sciences*, Daniel Graves]

**Louis Pasteur** (1822-1895) He is remembered for his remarkable breakthroughs in the causes and preventions of disease. His experiments supported the germ theory of disease. He is regarded as one of the three main founders of microbiology, together with Ferdinand Cohn and Robert Koch. Pasteur also made many discoveries in the field of chemistry, most notably the molecular basis for the asymmetry of certain crystals. Pasteur demonstrated that fermentation is caused by the growth of micro-organisms, and that the emergent growth of bacteria in nutrient broths is not due to spontaneous generation but rather to biogenesis (Omne vivum ex ovo). While Pasteur was not the first to propose germ theory (Girolamo Fracastoro, Agostino Bassi, Friedrich Henle and others had suggested it earlier), he developed it and conducted experiments that clearly indicated its correctness and managed to convince most of Europe it was true. Pasteur's research also showed that the growth of micro-organisms was responsible for spoiling beverages, such as beer, wine and milk. With this established, he invented a process (pasteurization) in which liquids such as milk were heated to kill most bacteria and molds already present within them. Beverage contamination led Pasteur to the idea that micro-organisms infecting animals and humans cause disease. He proposed preventing the entry of microorganisms into the human body, leading Joseph Lister to develop antiseptic methods in surgery. Pasteur also discovered anaerobiosis, whereby some microorganisms can develop and live without air or oxygen, called the Pasteur effect. The difference between the earlier smallpox vaccination and anthrax or chicken cholera vaccination was that the weakened form of the latter two disease organisms had been *generated artificially*, and so a naturally weak form of the disease organism did not need to be found. This discovery revolutionized work in infectious diseases, and Pasteur gave these artificially weakened diseases the generic name of *vaccines*. Pasteur produced the first vaccine for rabies by growing the virus in rabbits, and then weakening it by drying the affected nerve tissue. His discoveries also reduced mortality from puerperal fever. Catholic observers often said that Louis Pasteur remained through out his whole life an ardent Christian. His son-in-law, in perhaps the most complete biography of Louis Pasteur, writes:

> Absolute faith in God and in Eternity, and a conviction that the power for good given to us in this world will be continued beyond it, were feelings which pervaded his whole life; the virtues of the gospel had ever been present to him. Full of respect for the form of religion which had been that of his forefathers, he came simply to it and naturally for spiritual help in these last weeks of his life.

He died while listening to the story of St Vincent de Paul, whom he admired and sought to emulate. [source: Wikipedia bio] Although not a churchgoer, he was a Franciscan Tertiary and detested atheists and atheism. [source: *Christian Influences in the Sciences*, Daniel Graves] Pasteur stated, for example:

> The more I study nature, the more I stand amazed at the work of the Creator. Science brings men nearer to God.

> In good philosophy, the word cause ought to be reserved to the single Divine impulse that has formed the universe.

> Little science takes you away from God but more of it takes you to Him.

> [source: Tihomir Dimitrov, editor, *50 Nobel Laureates and Other Great Scientists Who Believe in God*]

**William Kitchen Parker** (1823-1890) Physician, zoologist and comparative anatomist. The honours and appointments Parker gained later in life were due mainly for his work on the vertebrate skeleton and its significance in establishing a "true theory of the vertebrate skull" (Edward Sabine). He developed about 300 preparations of bird wings and many complete skeletons. This work resulted in 24 papers on birds. From 1865 to 1888 Parker published 36 studies on the vertebrate skull, including a monograph. The entire series comprises nearly 1800 pages of letterpress and about 270 plates. [source: Wikipedia bio]

**Jean Henri Fabre** (1823-1915) He was a popular teacher, physicist, chemist and botanist, but best known for his findings in the field of entomology, the study of insects. He is considered by many to be the father of modern entomology. Fabre's influence is felt in the later works of fellow naturalist Charles Darwin, who called Fabre "an inimitable observer". Fabre, however, remained sceptical about Darwin's theory of evolution, as he always restrained from all theories and systems. His special force was exact and detailed observation, or field research, as we would call it today. Always opposed to atheism, he was a convert to committed Christianity later in his life. [source: Wikipedia bio]

**Pierre Jules Janssen** (1824-1907) Along with the English scientist Joseph Norman Lockyer, he is credited with discovering the gas helium. While observing a solar eclipse in 1868, he noticed a bright yellow line with a wavelength of 587.49 nm in the spectrum of the chromosphere of the Sun. This was the first observation of this particular spectral line, and one possible source for it was an element not yet discovered on the earth. Janssen was at first ridiculed since no element had ever been detected in space before being found on Earth. [source: Wikipedia bio]

**Lord Kelvin** (William Thomson) (1824-1907) He did important work in the mathematical analysis of electricity and formulation of the first and second Laws of Thermodynamics, and did much to unify the emerging discipline of physics in its modern form. He also had a career as an electric telegraph engineer and inventor. He made important improvements to the mariner's compass and developed the basis of absolute zero. He invented the mirror galvanometer and the siphon recorder, introduced a method of deep-sea sounding, revived the Sumner method of finding a ship's place at sea, and developed a tide predicting machine. He helped lay the first transatlantic cable and invented the current balance, also known as the *Kelvin balance* or *Ampere balance* (*SiC*), for the precise specification of the ampere, the standard unit of electric current. In 1900, he gave a lecture titled *Nineteenth-Century Clouds over the Dynamical Theory of Heat and Light*. The two "dark clouds" he was alluding to were the unsatisfactory explanations that the physics of the time could give for two phenomena: the Michelson–Morley experiment and black body radiation. Two major physical theories were developed during the twentieth century starting from these issues: for the former, the Theory of relativity; for the second, quantum mechanics. Lord Kelvin remained a devout believer in Christianity throughout his life: attendance at chapel was part of his daily routine. He saw his Christian faith as supporting and informing his scientific work. He was an elder of St Columba's Parish Church (Church of Scotland) in Largs for many years. [source: Wikipedia bio] Kelvin closed his presidential address to the British Association for the Advancement of Science (Edinburgh, August 1871) thus:

> Overpoweringly strong proofs of intelligent and benevolent design lie all around us; and if ever perplexities, whether metaphysical or scientific, turn us away from them for a time, they come back upon us with irresistible force, showing to us through Nature the influence of a free will, and teaching us that all living things depend on one ever-acting Creator and Ruler.

In his first lecture in the "Introductory Course of Natural Philosophy," he stated:

> We feel that the power of investigating the laws established by the Creator for maintaining the harmony and permanence of His works is the noblest privilege which He has granted to our intellectual state. As the

depth of our insight into the wonderful works of God increases, the stronger are our feelings of awe and veneration in contemplating them and in endeavoring to approach their Author.

In a speech to University College (1903), Kelvin said: "Do not be afraid to be free thinkers. If you think strongly enough, you will be forced by science to the belief in God." He once said, "The atheistic idea is so nonsensical that I cannot put it into words." In his address at the annual meeting of the Christian Evidence Society (May 23, 1889), Kelvin said:

> I have long felt that there was a general impression in the non-scientific world, that the scientific world believes Science has discovered ways of explaining all the facts of Nature without adopting any definite belief in a Creator. I have never doubted that that impression was utterly groundless.

He opposed materialism:

> Science can do little positively towards the objects of this society. But it can do something, and that something is vital and fundamental. It is to show that what we see in the world of dead matter and of life around us is not a result of the fortuitous concourse of atoms.

[source: Tihomir Dimitrov, editor, *50 Nobel Laureates and Other Great Scientists Who Believe in God*]

**Sir William Huggins** (1824-1910) He was best known for his pioneering work in astronomical spectroscopy. Huggins was the first to distinguish between nebulae and galaxies by showing that some (like the Orion Nebula) had pure emission spectra characteristic of gas, while others like the Andromeda Galaxy had spectra characteristic of stars. He was also the first to adopt dry plate photography in imaging astronomical objects. [source: Wikipedia bio]

**Blessed Francesco Faà di Bruno** (1825-1888) He made numerous and important contributions to mathematics. Today, he is best known for Faà di Bruno's formula on derivatives of composite functions. He was the author of an exhaustive treatise on the theory and applications of elliptic functions, in three volumes; *Théorie générale de l'élimination* (1859); *Calcolo degli errori* (1867), and most important of all, *Théorie des formes binaires* (1876). [source: Wikipedia bio]

**Thomas Henry Huxley** (1825-1895; agnostic and assuredly *not* an atheist) He became perhaps the finest comparative anatomist of the latter 19th century, working on invertebrates and clarifying relationships between groups previously little understood. Huxley was slow to accept some of Darwin's ideas, such as gradualism, and was undecided about natural selection. The correctness of natural selection as the main mechanism for evolution was to lie permanently in Huxley's mental pending tray. He never conclusively made up his mind about it, though he did admit it was a hypothesis which was a good working basis. His agnostic-type reservations on natural selection were of the type "until selection and breeding can be seen to give rise to varieties which are infertile with each other, natural selection cannot be proved". One reason for this doubt was that comparative anatomy could address the question of descent, but not the question of mechanism. Despite this he was wholehearted in his public support of Darwin, as his "bulldog" in debates. [source: Wikipedia bio] He wrote around 1859, concerning Darwin's theory of evolution:

> I was not brought into serious contact with the "Species" question until after 1850. . . . It seemed to me then (as it does now) that "creation," in the ordinary sense of the word, is perfectly conceivable. I find no difficulty in conceiving that, at some former period, this universe was not in existence; and that it made its appearance in six days (or instantaneously, if that is preferred), in consequence of the volition of some pre-existing Being. Then, as now, the so-called a priori arguments against

Theism, and, given a Deity, against the possibility of creative acts, appeared to me to be devoid of reasonable foundation. I had not then, and I have not now, the smallest a priori objection to raise to the account of the creation of animals and plants given in Paradise Lost, in which Milton so vividly embodies the natural sense of Genesis. Far be it from me to say that it is untrue because it is impossible. I confine myself to what must be regarded as a modest and reasonable request for some particle of evidence that the existing species of animals and plants did originate in that way, as a condition of my belief in a statement which appears to me to be highly improbable.

(*Life and letters of Thomas Henry Huxley, Volume 1*, edited by Leonard Huxley [New York: D. Appleton & Co., 1901], pp. 179-180)

Writing to Charles Kingsley on 6 May 1863, he opined:

I am too much a believer in Butler and in the great principle of the "Analogy" that "there is no absurdity in theology so great that you cannot parallel it by a greater absurdity of Nature" (it is not commonly stated in this way), to have any difficulties about miracles. I have never had the least sympathy with the a priori reasons against orthodoxy, and I have by nature and disposition the greatest possible antipathy to all the atheistic and infidel school.

Nevertheless, I know that I am, in spite of myself, exactly what the Christian world call, and, so far as I can see, are justified in calling, atheist and infidel. I cannot see one shadow or tittle of evidence that the great unknown underlying the phenomena of the universe stands to us in the relation of a Father — loves us and cares for us as Christianity asserts. On the contrary, the whole teaching of experience seems to me to show that while the governance (if I may use the term) of the universe is rigorously just and substantially kind and beneficent, there is no more relation of affection between governor and governed than between me and the twelve judges.

(*Ibid.*, pp. 259-260)

In 1869, he explained his self-description as an "agnostic":

When I reached intellectual maturity, and began to ask myself whether I was an atheist, a theist, or a pantheist; a materialist or an idealist; a Christian or a freethinker; I found that the more I learned and reflected, the less ready was the answer; until, at last, I came to the conclusion that I had neither art nor part with any of these denominations, except the last The one thing in which most of these good people were agreed was the one thing in which I differed from them. They were quite sure they had attained a certain "gnosis" — had, more or less successfully, solved the problem of existence; while I was quite sure I had not, and had a pretty strong conviction that the problem was insoluble. And, with Hume and Kant on my side, I could not think myself presumptuous in holding fast by that opinion. . . .

(*Ibid.*, p. 343)

In his 1874 essay, *On the Hypothesis That Animals Are Automata, and Its History*, he remarked:

. . . not among materialists, for I am utterly incapable of conceiving the existence of matter if there is no mind in which to picture that existence; not among atheists, for the problem of the ultimate cause of existence is one which seems to me to be hopelessly out of reach of my poor powers. Of all the senseless babble I have ever had occasion to read, the demonstrations of these philosophers who undertake to tell us all about the nature of God would be the worst, if they were not surpassed by the still greater absurdities of the philosophers who try to prove that there is no God.

(*Life and letters of Thomas Henry Huxley, Volume 2*, edited by Leonard Huxley [London: Macmillan, 2nd ed., 1903], p. 133; see also *Collected Essays, Vol. 1*, p. 245)

In his article, "Science and Morals" (1886), Huxley stated:

> The student of nature, who starts from the axiom of the universality of the law of causation, cannot refuse to admit an eternal existence; if he admits the conservation of energy, he cannot deny the possibility of an eternal energy; if he admits the existence of immaterial phenomena in the form of consciousness, he must admit the possibility, at any rate, of an eternal series of such phenomena; and, if his studies have not been barren of the best fruit of the investigation of nature, he will have enough sense to see that when Spinoza says, 'Per Deum intelligo ens absolute infinitum, hoc est substantiam constantem infinitis attributis,' the God so conceived is one that only a very great fool would deny, even in his heart. Physical science is as little Atheistic as it is Materialistic.

In "On Providence" (*An Apologetic Irenicon*, 1892), Huxley wrote:

> If the doctrine of a Providence is to be taken as the expression, in a way 'to be understanded of the people,' of the total exclusion of chance from a place even in the most insignificant corner of Nature; if it means the strong conviction that the cosmic process is rational; and the faith that, throughout all duration, unbroken order has reigned in the universe – I not only accept it, but I am

disposed to think it the most important of all truths. As it is of more consequence for a citizen to know the law than to be personally acquainted with the features of those who will surely carry it into effect, so this very positive doctrine of Providence, in the sense defined, seems to me far more important than all the theorems of speculative theology.

[source: Tihomir Dimitrov, editor, *50 Nobel Laureates and Other Great Scientists Who Believe in God*]

Bernhard Riemann (1826-1866) In school, the Lutheran Riemann had tried to prove mathematically the correctness of the Book of Genesis. At age 19, he started studying theology. Riemann held his first lectures in 1854, which founded the field of Riemannian geometry and thereby set the stage for Einstein's general theory of relativity. He was also the first to suggest using dimensions higher than merely three or four in order to describe physical reality. Riemann's published works opened up research areas combining analysis with geometry. These would subsequently become major parts of the theories of Riemannian geometry, algebraic geometry, and complex manifold theory. The theory of Riemann surfaces was elaborated by Felix Klein and particularly Adolf Hurwitz. This area of mathematics is part of the foundation of topology, and is still being applied in novel ways to mathematical physics. He made major contributions to real analysis. He defined the Riemann integral by means of Riemann sums, developed a theory of trigonometric series that are not Fourier series—a first step in generalized function theory—and studied the Riemann–Liouville differintegral. In the area of modern analytic number theory, he introduced the Riemann zeta function and established its importance for understanding the distribution of prime numbers. He made a series of conjectures about properties of the zeta function, one of which is the well-known Riemann hypothesis. He applied the Dirichlet principle from variational calculus to great effect, and his work on monodromy and the hypergeometric function in the complex domain was also influential. His theory of higher dimensions ("On the hypotheses which underlie geometry") was received with enthusiasm, and it is one of the most important works in geometry. He introduced a collection of numbers at every point in space (i.e., a tensor) which would describe how much it was bent or curved. Riemann found that in four spatial dimensions, one needs a collection of ten numbers at each point to describe the properties of a manifold, no matter how distorted it is. This is the famous construction central to his geometry, known now as a Riemannian metric. He was a devout Christian. [source: Wikipedia bio]

**Armand David** (1826-1900; priest) Lazarist missionary and zoologist and a botanist. He succeeded in obtaining many specimens of hitherto unknown animals and plants, and the value of his comprehensive collections for the advance of systematic zoology and especially for the advancement of animal geography received universal recognition from the scientific world. He found in China 200 species of wild animals, of which 63 were hitherto unknown to zoologists, and 807 species of birds, 65 of which had not been described before. He made a large collection of reptiles, batrachians, and fishes and handed it over to specialists for further study; also a large number of moths and insects, many of them hitherto unknown. Among the rhododendrons which he collected no less than fifty-two new species were found and among the primulae about forty. The most remarkable of the animals found by David which were hitherto-unknown to Europe were the Giant Panda and Père David's Deer. [source: Wikipedia bio]

**Joseph Lister** (1827-1912) He was a surgeon and a pioneer of antiseptic surgery, who promoted the idea of sterile surgery. Lister successfully introduced carbolic acid (now called phenol) to sterilize surgical instruments and to clean wounds, which led to reduced post-operative infections and made surgery safer for patients. Until Lister's studies on antiseptics, most people believed that chemical damage from exposure to bad air (see "miasma") was responsible for infections in wounds. Lister also noticed that midwife-delivered babies had a lower mortality rate than surgeon-delivered babies, correctly attributing this difference to the fact that midwives tended to wash their hands more often than surgeons, and that surgeons often would go directly from one surgery, such as draining an abscess, to delivering a baby. He instructed surgeons under his responsibility to wear clean gloves and wash their hands before and after operations with 5% carbolic acid solutions. Instruments were also washed in the same solution and assistants sprayed the solution in the operating theatre. One of his additional suggestions was to stop using porous natural materials in manufacturing the handles of medical instruments. As the germ theory of disease became more widely accepted, it was realised that infection could be better avoided by preventing bacteria from getting into wounds in the first place. This led to the rise of sterile surgery. Some consider Lister "the father of modern antisepsis". He was the second man in England to operate on a brain tumor; he developed a method of repairing kneecaps with metal wire, and improved the technique of mastectomy. Lister was said to be a shy, unassuming man, deeply religious in his beliefs (Episcopalian), and uninterested in social success or financial gain. [source: Wikipedia bio] Lister was reared a devout Quaker and migrated to the Church of England. He reminded his pupils that they had to be prepared to give an account to God for their treatment of "the earthly tabernacle" of the soul (ie: the human body). [source: *Christian Influences in the Sciences*, Daniel Graves]

**Sir Benjamin Ward Richardson** (1828-1896) He was a physician, anaesthetist, physiologist, sanitarian, and prolific writer on medical history. He was a committed exponent of the microbial cause of infectious disease, brought into clinical use, no less than fourteen anesthetics, of which methylene bichloride is the best known, and invented the first double-valved mouthpiece for use in the administration of chloroform. He also produced local insensibility by freezing the part with an ether spray. He was also one of the earliest advocates of bicycling (in 1883). [source: Wikipedia bio]

George Rolleston (1829-1881) Physician and zoologist. His research included comparative anatomy, physiology, zoology, archaeology, and anthropology. He held the positions of Lee's Reader in Anatomy, Christ Church, Oxford (1857) and Linacre Professor of Anatomy and Physiology, Oxford (1860). Rolleston was an Anglican. He once remarked that whenever he lectured on evolution, he was asked "Was I an atheist or a Unitarian?" [source: Wikipedia bio]

James Clerk Maxwell (1831-1879) His most important achievement was classical electromagnetic theory, synthesizing all previously unrelated observations, experiments and equations of electricity, magnetism and even optics into a consistent theory. His set of equations—Maxwell's equations—demonstrated that electricity, magnetism and even light are all manifestations of the same phenomenon: the electromagnetic field. From that moment on, all other classic laws or equations of these disciplines became simplified cases of Maxwell's equations. Maxwell's work in electromagnetism has been called the "second great unification in physics", after the first one carried out by Isaac Newton. Maxwell demonstrated that electric and magnetic fields travel through space in the form of waves, and at the constant speed of light. He proposed that light was in fact undulations in the same medium that is the cause of electric and magnetic phenomena. His work in producing a unified model of electromagnetism is considered to be one of the greatest advances in physics. Maxwell also developed the Maxwell–Boltzmann distribution, a statistical means of describing aspects of the kinetic theory of gases. These two discoveries helped usher in the era of modern physics, laying the foundation for future work in such fields as special relativity and quantum mechanics. Maxwell is also known for creating the first true colour photograph in 1861 and for his foundational work on the rigidity of rod-and-joint frameworks like those in many bridges. In the end of millennium poll, a survey of the 100 most prominent physicists, Maxwell was voted the third greatest physicist of all time, behind only Newton and Einstein. Einstein kept a photograph of Maxwell on his study wall, alongside pictures of Michael Faraday and Isaac Newton, and he stated:

"The special theory of relativity owes its origins to Maxwell's equations of the electromagnetic field." Also, "Since Maxwell's time, physical reality has been thought of as represented by continuous fields, and not capable of any mechanical interpretation. This change in the conception of reality is the most profound and the most fruitful that physics has experienced since the time of Newton." Ivan Tolstoy, author of one of Maxwell's biographies, remarked at the frequency with which scientists writing short biographies on Maxwell often omit the subject of his Christianity. Maxwell's religious beliefs and related activities have been the focus of several peer-reviewed and well-referenced papers. Attending both Presbyterian and Episcopalian services as a child, Maxwell later underwent an evangelical conversion (April 1853). [source: Wikipedia bio] For more on his religious views, see: Jerrold L. McNatt, "James Clerk Maxwell's Refusal to Join the Victoria Institute" (PDF), Ian Hutchinson, "James Clerk Maxwell and the Christian Proposition," and Paul Theerman, "James Clerk Maxwell and Religion," *American Journal of Physics*, Vol. 54, Issue 4 (April 1986).

**Sir William Henry Flower** (1831-1899) Comparative anatomist and surgeon. Flower became a leading authority on mammals, and especially on the primate brain. He also became an expert on the Cetacea (whales and their relatives). He carried out dissections, went out on whaling boats, arranged whale exhibits and studied the new discoveries of whale fossils. He made valuable contributions to structural anthropology, publishing, for example, complete and accurate measurements of no less than 1,300 human skulls. He studied marsupials as well, and was the first person to show that lemurs are primates. He combined religious belief with an unequivocal acceptance of evolution. His point of view was close to that of Asa Gray, the American botanist, who wrote a pamphlet entitled *Natural Selection not inconsistent with Natural Theology*. As the years passed this co-existence of ideas became ever more common. In 1883 Flower gave an address to the Church Congress in Reading on evolution: "The bearing of science on religion" (reprinted in his *Essays on Museums*). [source: Wikipedia bio]

**James Bell Pettigrew** (1832-1908) Distinguished naturalist. In 1860, he advanced a remarkable discussion of the anatomy of the musculature of the heart. From an early age, Pettigrew demonstrated a remarkable gift for morphological analysis and an analytical grasp of natural history. In 1873 he published *Animal Locomotion: or Walking, Swimming and Flying*, his most popular work. Later he assembled his magnum opus, *Design in Nature*, published in three volumes and lavishly illustrated with engravings and photographs. In this work, he showed himself to be almost indifferent in relation to Darwinism. [source: Wikipedia bio]

**Dmitri Mendeleev** (1834-1907) Creator of the first version of the periodic table of elements. Using the table, he predicted the properties of elements yet to be discovered. Mendeleev wrote the definitive two-volume textbook at that time: *Principles of Chemistry* (1868–1870). As he attempted to classify the elements according to their chemical properties, he noticed patterns that led him to postulate his Periodic Table. Mendeleev devoted much study and made important contributions to the determination of the nature of such indefinite compounds as solutions. In another department of physical chemistry, he investigated the expansion of liquids with heat, and devised a formula similar to Gay-Lussac's law of the uniformity of the expansion of gases, while in 1861 he anticipated Thomas Andrews' conception of the critical temperature of gases by defining the absolute boiling-point of a substance as the temperature at which cohesion and heat of vaporization become equal to zero and the liquid changes to vapor, irrespective of the pressure and volume. He studied petroleum origin and concluded hydrocarbons are abiogenic and form deep within the earth - see Abiogenic petroleum origin. [source: Wikipedia bio]

**Giovanni Schiaparelli** (1835-1910) He was an astronomer, particularly known for his studies of Mars. In his initial observations, he named the "seas" and "continents" of Mars. During the planet's "Great Opposition" of 1877, he observed a dense network of linear structures on the surface of Mars which he called *canali* in Italian, meaning *channels* but the term was mistranslated into English as "canals". While the latter term indicates an artificial construction, the former indicates the connotation that it can also be a natural configuration of the land. Schiaparelli demonstrated that the Perseid and Leonid meteor showers were associated with comets. He proved, for example, that the orbit of the Leonids coincided with that of the Comet Tempel-Tuttle. These observations led the astronomer to formulate the hypothesis, subsequently proved to be very exact, that the meteor showers could be the trails of comets. [source: Wikipedia bio]

**Jean Baptiste Carnoy** (1836–1899; priest) He was the founder of the science of cytology (cell biology). He made the initial explanation of the real nature of the albuminoid membrane, and conducted noted experiments on cellular segmentation. [source: Wikipedia bio]

**Josiah Willard Gibbs** (1839-1903) He devised much of the theoretical foundation for chemical thermodynamics as well as physical chemistry. As a mathematician, he invented vector analysis. In 1901, Gibbs was awarded the highest possible honor granted by the international scientific community of his day: the Copley Medal of the Royal Society of London, for his greatest contribution, that being "the first to apply the second law of thermodynamics to the exhaustive discussion of the relation between chemical, electrical, and thermal energy and capacity for external work." Gibbs applied thermodynamics to interpret physicochemical phenomena, successfully explaining and interrelating what had previously been a mass of isolated facts. He also wrote on optics, and developed a new electrical theory of light. His chemical thermodynamics was a theory of greater generality than any other theory of matter extant in his day. After 1889, he worked on statistical mechanics, laying a foundation and "providing a mathematical framework for quantum theory and for Maxwell's theories". He wrote classic textbooks on statistical mechanics. Gibbs also contributed to crystallography and applied his vector methods to the determination of planetary and comet orbits. Max Planck said that Gibbs "will ever be reckoned among the most renowned theoretical physicists of all times." [source: Wikipedia bio] He was a man of quiet Christianity who showed it in conscientious work and steady churchgoing. [source: *Christian Influences in the Sciences*, Daniel Graves]

**John William Strutt (Lord Rayleigh)** (1842-1919) With William Ramsay, he discovered the element argon. He also discovered in 1871 the phenomenon now called Rayleigh scattering, explaining why the sky is blue, and predicted the existence of the surface waves now known as Rayleigh waves. [source: Wikipedia bio] He wrote his classic *The Theory of Sound* in two volumes (1877-1878). The first volume dealt with the mechanics of a vibrating medium which produces sound; the second volume was about acoustic wave propagation. He standardized the ohm. He wrote a remarkable 446 publications, covering an incredible range of topics in applied mathematics and physics. In 1879 Rayleigh wrote a paper on traveling waves, this theory has now developed into the theory of solitons, and developed the Rayleigh-wave theory in 1885. He contributed to hydrodynamics: in particular to hydrodynamic similarity. [source: MacTutor Archive]

**Georg Cantor** (1845-1918) Mathematician, best known as the inventor (between 1874 and 1884) of set theory, which has become a fundamental theory in mathematics. Cantor established the importance of one-to-one correspondence between sets, defined infinite and well-ordered sets, and proved that the real numbers are "more numerous" than the natural numbers (see Cantor–Bernstein–Schroeder theorem). In fact, Cantor's theorem implies the existence of an "infinity of infinities". He defined the cardinal and ordinal numbers and their arithmetic. Other contributions include his famous diagonal argument and the Cantor set. He was also the first to formulate what later came to be known as the continuum hypothesis. Cantor believed his theory of transfinite numbers had been communicated to him by

God, who had chosen Cantor to reveal them to the world. To Cantor, his mathematical views were intrinsically linked to their philosophical and theological implications—he identified the Absolute Infinite with God. He stated that "the transfinite species are just as much at the disposal of the intentions of the Creator and His absolute boundless will as are the finite numbers." Cantor also believed that his theory of transfinite numbers ran counter to both materialism and determinism. Though Lutheran, he even sent one letter directly to Pope Leo XIII himself, and addressed several pamphlets to him. [source: Wikipedia bio]

**Alexander Graham Bell** (1847-1922; Unitarian) He was an eminent scientist, inventor, engineer and innovator who is credited with inventing the first practical telephone (1876). His other patents included four for the Photophone, one for the phonograph, five for aerial vehicles, four for "hydroairplanes" and two for selenium cells. His many other inventions included groundbreaking work in optical telecommunications, hydrofoils and aeronautics, a metal jacket to assist in breathing, the audiometer to detect minor hearing problems, a device to locate icebergs, investigations on how to separate salt from seawater, and work on finding alternative fuels. Bell worked extensively in medical research and invented techniques for teaching speech to the deaf. He considered impressing a magnetic field on a record as a means of reproducing sound, but was unable to develop a workable prototype. Bell had glimpsed a basic principle which would one day find its application in the tape recorder, the hard disc and floppy disc drive and other magnetic media. His own home used a primitive form of air conditioning, in which fans blew currents of air across great blocks of ice. He reflected on the possibility of using solar panels to heat houses. Bell is also credited with the invention of the metal detector in 1881. He developed the first practical hydrofoil watercraft. [source: Wikipedia bio]

**Sir William Ramsay** (1852-1916) Discovered the noble gases and received the Nobel Prize in Chemistry in 1904 "in recognition of his services in the discovery of the inert gaseous elements in air." In 1894 he isolated a heavy component of air previously unknown that did not appear to have any obvious chemical reactivity. He named the gas "argon". In the years that followed he discovered neon, krypton, and xenon. He also isolated helium which had been observed in the spectrum of the sun but had not been found on earth. In 1910 he also made and characterized radon. [source: Wikipedia bio]

**Aleksey Pavlovich Hansky** (1870-1908) He undertook research on the form of the solar corona in relation to the phases of solar activity. Hansky studied gravimetry, measuring gravitational force atop Mont Blanc and in the depths of a coal mine on Spitsbergen, conducted research on the zodiacal light, and made observations of Jupiter and the meteors. But his chief service to science was his solar research. He compared his photographs of the solar corona taken during the 1896 eclipse with data on the various phenomena of solar activity over a fifty-year period, and postulated a relation between the form of the solar corona and the number of sunspots, that is, with the phase of solar activity. It appeared that when there is a minimum of spots, the corona is stretched along the plane of the solar equator but is scarcely observable at the poles, and its total luminosity is only slightly greater than the luminosity of the full moon. But during the period of maximum sunspots the corona is ten times brighter than the full moon and is rather evenly distributed on all sides of the solar disk. During later eclipses Hansky's predictions of the form of the corona were fully confirmed. Hansky had striking success in photographing sunspots and details of solar granulation; only in the 1960s, with the launching of telescopes on special balloons, have better photographs been obtained. [source: Encyclopedia.com bio]

## Chapter Eight

## 31 Catholic, Protestant and Otherwise Religious Prominent Scientists: 1900-1950 (From Einstein to Planck, Eddington, and Lemaître)

Thomas Alva Edison (1847-1931) Inventor and scientist who developed many devices, including the phonograph, the motion picture camera, and a long-lasting, practical electric light bulb. He was one of the first inventors to apply the principles of mass production and large teamwork to the process of invention, and therefore is often credited with the creation of the first industrial research laboratory. Edison is considered one of the most prolific inventors in history, holding 1,093 U.S. patents in his name, as well as many patents in the United Kingdom, France, and Germany. He is credited with numerous inventions that contributed to mass communication and, in particular, telecommunications. These included a stock ticker, a mechanical vote recorder, a battery for an electric car, electrical power, recorded music and motion pictures. Edison originated the concept and implementation of electric-power generation and distribution to homes, businesses, and factories – a crucial development in the modern industrialized world. In 1877–1878, Edison invented and developed the carbon microphone used in all telephones along with the Bell receiver until the 1980s. Edison is credited with designing and producing the first commercially available fluoroscope, a machine that uses X-rays to take radiographs. Edison was heavily influenced by Thomas Paine's *The Age of Reason*. Edison defended Paine's "scientific deism", saying, "He has been called an atheist, but atheist he was not. Paine believed in a supreme intelligence, as representing the idea which other men often express by the name of deity." He wrote in 1910: "what you call God I call Nature, the Supreme intelligence that rules matter." [source: Wikipedia bio]

**Sir John Ambrose Fleming** (1849-1945) Electrical engineer and physicist known for inventing the first thermionic valve or vacuum tube, the diode, in 1904. He also invented the right-hand rule, used in mathematics and electronics. In 1892, Fleming presented an important paper on electrical transformer theory. In November 1904, he invented the two-electrode vacuum-tube rectifier, which he called the oscillation valve. It was also called a thermionic valve, vacuum diode, kenotron, thermionic tube, or Fleming valve. This invention is often considered to have been the beginning of electronics, for this was the first vacuum tube. Fleming's diode was used in radio receivers and radars for many decades afterwards, until it was superseded by solid state electronic technology more than 50 years later. He was an early advocate of the new technology of television. Fleming also contributed in the fields of photometry, electronics, wireless telegraphy (radio), and electrical measurements. He was a devout Christian and preached on one occasion at St Martin-in-the-Fields in London on the topic of evidence for the resurrection. In 1932, along with Douglas Dewar and Bernard Acworth, he helped establish the Evolution Protest Movement. Having no children, he bequeathed much of his estate to Christian charities, especially those that helped the poor. [source: Wikipedia bio]

**Sir Joseph John "J. J." Thomson** (1856-1940) Physicist who is credited for the discovery of the electron and of isotopes, and the invention of the mass spectrometer. Thomson was awarded the 1906 Nobel Prize in Physics for the discovery of the electron and for his work on the conduction of electricity in gases. Thomson discovered the electron through his explorations on the properties of cathode rays. He found that the rays could be deflected by an electric field (in addition to magnetic fields, which was already known). He concluded that these rays, rather than being waves, were composed of very light negatively charged particles which he called "corpuscles". (Later scientists preferred the name electron which had been suggested by George Johnstone Stoney in 1894, prior to Thomson's actual discovery). Thomson believed that the corpuscles emerged from the atoms of the trace gas inside his cathode ray tubes. He thus concluded that atoms were divisible, and that the corpuscles were their building blocks. In 1905 Thomson discovered the natural radioactivity of potassium and in 1906 demonstrated that hydrogen had only a single electron per atom. [source: Wikipedia bio] His inaugural presidential address to the British Association was published in the prominent scientific journal *Nature* (26 August 1909, vol. 81, p. 257). Thomson concluded his address with the following words:

> As we conquer peak after peak we see in front of us regions full of interest and beauty, but we do not see our goal, we do not see the horizon; in the distance tower still higher peaks, which will yield to those who ascend them still wider prospects, and deepen the feeling, the truth of which is emphasized by every advance in science, that 'Great are the Works of the Lord'.
>
> [source: excerpt from Tihomir Dimitrov, editor, *50 Nobel Laureates and Other Great Scientists Who Believe in God* (online book, 2008) ]

Max Planck (1858-1947) Physicist who is considered to be the founder of the quantum theory, and thus one of the most important physicists of the twentieth century. In 1894 Planck turned his attention to the problem of black-body radiation. The problem had been stated by Kirchhoff in 1859: how does the intensity of the electromagnetic radiation emitted by a black body (a perfect absorber, also known as a cavity radiator) depend on the frequency of the radiation (i.e., the color of the light) and the temperature of the body? By 1900 he had derived the first version of the famous Planck black-body radiation law, which described the experimentally observed black-body spectrum. In November 1900, Planck revised this first approach, relying on Boltzmann's statistical interpretation of the second law of thermodynamics as a way of gaining a more fundamental understanding of the principles behind his radiation law. As Planck was deeply suspicious of the philosophical and physical implications of such an interpretation of Boltzmann's approach, his recourse to them was, as he later put it, "an act of despair ... I was ready to sacrifice any of my previous convictions about physics." His central assumptions were formulated in his Planck postulate, that electromagnetic energy could be emitted only in quantized form, and Planck's constant. At first Planck considered that quantisation was only "a purely formal assumption ... actually I did not think much about it."; nowadays this assumption, incompatible with classical physics, is regarded as the birth of quantum physics. The discovery of Planck's constant enabled him to define a new universal set of physical units (such as the Planck length and the Planck mass), all based on fundamental physical constants. Planck was among the few who immediately recognized the significance of Einstein's special theory of relativity. Thanks to his influence this theory was soon widely accepted in Germany. Planck also contributed considerably to extend the special theory of relativity. At the end of the 1920s Bohr, Heisenberg and Pauli had worked out the Copenhagen interpretation of quantum mechanics, but it was rejected by Planck, as well as Schrödinger, Laue, and Einstein. Planck experienced the truth of his own earlier observation from his struggle with the older views in his younger years: "A new scientific truth does not triumph by convincing its opponents and making them see the light, but rather because its opponents eventually die, and a new generation grows up that is familiar with it." Planck was a devoted and persistent adherent of Christianity from early life to death. The God in which Planck believed was an almighty, all-knowing, benevolent but unintelligible God that permeated everything, manifest through symbols, including physical laws. He regarded the scientist as a man of imagination and faith, "faith" interpreted as being similar to "having a working hypothesis". For example the causality principle isn't true or false, it is an act of faith. [source: Wikipedia bio] In his famous May 1937 lecture *Religion and Science*, Planck asserted:

> Both religion and science need for their activities the belief in God, and moreover God stands for the former in the beginning, and for the latter at the end of the whole thinking. For the former, God represents the basis, for the latter – the crown of any reasoning concerning the worldview. . . .
> It is the steady, ongoing, never-slackening fight against scepticism and dogmatism, against unbelief and superstition, which religion and science wage together. The directing watchword in this struggle runs from the remotest past to the distant future: "On to God!"

(Max Planck, *Religion und Naturwissenschaft*, Leipzig: Johann Ambrosius Barth Verlag, 1958, pp. 27, 30)

In his book, *Where Is Science Going?* (1932) Planck pointed out:

> There can never be any real opposition between religion and science; for the one is the complement of the other. Every serious and reflective person realizes, I think, that the religious element in his nature must be recognized and cultivated if all the powers of the human soul are to act together in perfect balance and harmony. And indeed it was not by accident that the greatest thinkers of all ages were deeply religious souls.

(1977 reprint, p. 168)

[source: excerpt from Tihomir Dimitrov, editor, *50 Nobel Laureates and Other Great Scientists Who Believe in God* (online book, 2008) ]

**Pierre Maurice Marie Duhem** (1861-1916) Physicist, mathematician and historian and philosopher of science, best known for his writings on the indeterminacy of experimental criteria and on scientific development in the Middle Ages. Duhem also made major contributions to the science of his day, particularly in the fields of hydrodynamics, elasticity, and thermodynamics. Regarding the latter, he was partly responsible for the development of what is known as the Gibbs–Duhem relation and the Duhem–Margules equation. Duhem thought that from thermodynamics first principles physicists should be able to derive all the other fields of physics—e.g., chemistry, mechanics, and electromagnetism. [source: Wikipedia bio]

**Sir William Henry Bragg** (1862-1942) Physicist, chemist, and mathematician. In 1915 he and his son William Lawrence Bragg were jointly awarded the Nobel Prize in Physics for their studies, using the X-ray spectrometer, of X-ray spectra, X-ray diffraction, and of crystal structure. They invented the X-ray spectrometer and founded the new science of X-ray analysis of crystal structure. Ten years later, their volume *X-Rays and Crystal Structure* (1915) had reached a fifth edition. [source: Wikipedia bio]

**Wilbur Wright** (1867-1912) and **Orville Wright** (1871-1948) are generally credited with inventing and building the world's first successful airplane and making the first controlled, powered and sustained heavier-than-air human flight, on December 17, 1903. In the two years afterward, the brothers developed their flying machine into the first practical fixed-wing aircraft. Although not the first to build and fly experimental aircraft, the Wright brothers were the first to invent aircraft controls that made fixed-wing powered flight possible. The brothers' fundamental breakthrough was their invention of three-axis control, which enabled the pilot to steer the aircraft effectively and to maintain its equilibrium.This method became standard and remains standard on fixed-wing aircraft of all kinds. The Wright brothers focused on unlocking the secrets of control to conquer "the flying problem", rather than developing more powerful engines as some other experimenters did. Their careful wind tunnel tests produced better aeronautical data than any before, enabling them to design and build wings and propellers more effective than any before. [source: Wikipedia bio]

**Henrietta Swan Leavitt** (1868-1921) Astronomer. She noted thousands of variable stars in images of the Magellanic Clouds. In 1908 she published her results in the *Annals of the Astronomical Observatory of Harvard College*, noting that a few of the variables showed a pattern: brighter ones appeared to have longer periods. After further study, she confirmed in 1912 that the variable stars of greater intrinsic luminosity – actually Cepheid variables – did indeed have longer periods, and the relationship was quite close and predictable. Leavitt's discovery is known as the "period-luminosity relationship" and showed, as she wrote, "that there is a simple relation between the brightness of the variable and their periods." This relationship provided an important yardstick for measuring distances in the Universe, if it could be calibrated. One year after Leavitt reported her results, Ejnar Hertzsprung determined the distance of several Cepheids in the Milky Way, and with this calibration the distance to any Cepheid could be determined. Cepheids were soon detected in other galaxies such as the Andromeda Galaxy (notably by Edwin Hubble in 1923–24). Cepheids were an important part of the evidence that galaxies are far outside of the Milky Way. Our picture of the universe was changed forever, largely because of Leavitt's discovery. The accomplishments of Edwin Hubble, renowned American astronomer, were made possible by Leavitt's groundbreaking research and Leavitt's Law. Hubble himself often said that Leavitt deserved the Nobel for her work. [source: Wikipedia bio]

**Robert A. Millikan** (1868-1953) Experimental physicist, and Nobel laureate in physics for his measurement of the charge on the electron and for his work on the photoelectric effect. Starting in 1909, Millikan worked on an oil-drop experiment in which they measured the charge on a single electron. The elementary charge is one of the fundamental physical constants and accurate knowledge of its value is of great importance. His experiment measured the force on tiny charged droplets of oil suspended against gravity between two metal electrodes. Knowing the electric field, the charge on the droplet could be determined. The beauty of the oil-drop experiment is that as well as allowing quite accurate determination of the fundamental unit of charge, Millikan's apparatus also provided a "hands on" demonstration that charge is actually quantized. When Einstein published his seminal 1905 paper on the particle theory of light, Millikan was convinced that it had to be wrong. He undertook a decade-long experimental program to test Einstein's theory. His results confirmed Einstein's predictions in every detail, but at first Millikan was not convinced of Einstein's interpretation. In his 1958 Book of discoveries on science experiments, however, he simply declared that his work "scarcely permits of any other interpretation than that which Einstein had originally suggested, namely that of the semi-corpuscular or photon theory of light itself." Since Millikan's work formed some of the basis for modern particle physics, it is ironic that he was rather conservative in his opinions about 20th century developments in physics, as in the case of the photon theory. He is also credited with measuring the value of Planck's constant by using photoelectric emission graphs of various metals. Millikan thought "cosmic ray" photons were the "birth cries" of new atoms continually being created by God to counteract entropy and prevent the heat death of the universe. In his later life he became interested in the relationship between Christian faith and science, and authored the book, *Evolution in Science and Religion.* [source: Wikipedia bio] In his *Autobiography* (Chapter 21 "The Two Supreme Elements in Human Progress") Robert Millikan wrote:

> Human well-being and all human progress rest at bottom upon two pillars, the collapse of either one of which will

bring down the whole structure. These two pillars are the cultivation and the dissemination throughout mankind of 1) the spirit of religion, and 2) the spirit of science (or knowledge).

(New York: Prentice-Hall, Inc., 1950, p. 279)
In an interview, entitled "A Scientist's God" (*Collier's*; 24 October 1925) Millikan stated:

This much I can say with definiteness - namely, that there is no scientific basis for the denial of religion - nor is there in my judgment any excuse for a conflict between science and religion, for their fields are entirely different. Men who know very little of science and men who know very little of religion do indeed get to quarreling, and the onlookers imagine that there is a conflict between science and religion, whereas the conflict is only between two different species of ignorance. . . .

Many of our great scientists have actually been men of profound religious convictions and life: Sir Isaac Newton, Michael Faraday, James Clerk Maxwell, Louis Pasteur. All these men were not only religious men, but they were also faithful members of their communions. For the most important thing in the world is a belief in moral and spiritual values – a belief that there is a significance and a meaning to existence – a belief that we are going somewhere! These men could scarcely have been so great had they been lacking in this belief.

[source: excerpt from Tihomir Dimitrov, editor, *50 Nobel Laureates and Other Great Scientists Who Believe in God* (online book, 2008) ]

**Ernest Rutherford** (1871-1937) Chemist and physicist; known as the father of nuclear physics. In early work he discovered the concept of radioactive half life, proved that radioactivity involved the transmutation of one chemical element to another, and also differentiated and named alpha and beta radiation, based on the two distinct types of radiation emitted by thorium and uranium (in 1899). In 1903, Rutherford realized that a type of radiation from radium discovered (but not named) by French chemist Paul Villard in 1900, must represent something different from alpha rays and beta rays, due to its very much greater penetrating power. Rutherford gave this third type of radiation its name also: the gamma ray. He was awarded the Nobel Prize in Chemistry in 1908. Along with Hans Geiger and Ernest Marsden he carried out the Geiger–Marsden experiment in 1909, which demonstrated the nuclear nature of atoms. It was his interpretation of this experiment that led him to formulate the Rutherford model of the atom in 1911 — that a very small positively charged nucleus was orbited by electrons, through his discovery and interpretation of Rutherford scattering in his gold foil experiment. In Cambridge in 1919 he became the first person to transmute one element into another when he converted nitrogen into oxygen through the nuclear reaction $^{14}N + \alpha \rightarrow {}^{17}O + p$. In 1921, while working with Niels Bohr (who postulated that electrons moved in specific orbits), Rutherford theorized about the existence of neutrons, which could somehow compensate for the repelling effect of the positive charges of protons by causing an attractive nuclear force and thus keeping the nuclei from breaking apart. Rutherford's theory of neutrons was proved in 1932 by his associate James Chadwick. He is widely credited with first splitting the atom in 1917, and leading the first experiment to "split the nucleus" in a controlled manner by two students under his direction, John Cockcroft and Ernest Walton in 1932. [source: Wikipedia bio]

**Alexis Carrel** (1873-1944) Surgeon and biologist who was awarded the Nobel Prize in Physiology or Medicine in 1912. He developed new techniques in vascular sutures and was a pioneer in transplantology and thoracic surgery. He collaborated with American physician Charles Claude Guthrie in work on vascular suture and the transplantation of blood vessels and organs. In 1894 he set about developing new techniques for suturing blood vessels. The technique of "triangulation", which was inspired by sewing lessons he took from an embroideress, is still used today. Julius Comroe wrote: "Between 1901 and 1910, Alexis Carrel, using experimental animals, performed every feat and developed every technique known to vascular surgery today." He had great success in reconnecting arteries and veins, and performing surgical grafts. During World War I (1914-1918), Carrel and the English chemist Henry Drysdale Dakin developed the Carrel-Dakin method of treating wounds based on chlorine (Dakin's solution) which, preceding the development of antibiotics, was a major medical advance in the care of traumatic wounds. Carrel co-authored a book with famed pilot Charles A. Lindbergh, *The Culture of Organs*, and worked with Lindbergh in the mid-1930s to create the "perfusion pump," which allowed living organs to exist outside of the body during surgery. The advance is said to have been a crucial step in the development of open-heart surgery and organ transplants, and to have laid the groundwork for the artificial heart, which became a reality decades later. Alexis Carrel went from being a skeptic of the visions and miracles reported at Lourdes to being a believer after experiencing a healing he could not explain. He refused to discount a supernatural explanation and steadfastly reiterated his beliefs, even writing a book describing his experience: *The Voyage to Lourdes* (New York, Harper & Row, 1939). This was a detriment to his career and reputation among his fellow doctors, and feeling he had no future in academic medicine in France, he emigrated to Canada with the intention of farming and raising cattle. After a brief period, he accepted an appointment at the University of Chicago. [source: Wikipedia bio] He wrote, in obvious agreement:

> Millikan, Eddington, and Jeans believe, like Newton, that the cosmos is the product of a Creative Intelligence.
>
> (*Reflections on Life* [New York: Hawthorn Books, Inc.: 1952], Chap. 6, Part 6; source: excerpt from Tihomir Dimitrov, editor, *50 Nobel Laureates and Other Great Scientists Who Believe in God* [online book, 2008] )

**Edmund Taylor Whittaker** (1873-1956) Mathematician who contributed widely to applied mathematics, mathematical physics and the theory of special functions. He had a particular interest in numerical analysis, but also worked on celestial mechanics and the history of physics. Near the end of his career he received the Copley Medal, the most prestigious honorary award in British science, "for his distinguished contributions to both pure and applied mathematics and to theoretical physics". Whittaker is remembered as the author of *A Course of Modern Analysis* (1902), which in its 1915 second edition in collaboration with George Neville Watson became *Whittaker and Watson*, one of the handful of mathematics texts of its era to become indispensable. This work has remained in print continuously for over a century. Whittaker is the eponym of the Whittaker function or Whittaker integral, in the theory of confluent hypergeometric functions. This makes him also the eponym of the Whittaker model in the local theory of automorphic representations. He published also on algebraic functions and automorphic functions. He gave expressions for the Bessel functions as integrals involving Legendre functions. In the theory of partial differential equations, Whittaker developed a general solution of the Laplace equation in three dimensions and the solution of the wave equation. He developed the electrical potential field as a bi-directional flow of energy (sometimes referred to as alternating currents). Whittaker's pair of papers in 1903 and 1904 indicated that any potential can be analysed by a Fourier-like series of waves, such as a planet's gravitational field point-charge. The superpositions of inward and outward wave pairs produce the "static" fields (or scalar potential). These were harmonically-related. By this conception, the structure of electric potential is created from two opposite, though balanced, parts. Whittaker suggested that gravity possessed a wavelike "undulatory" character. Whittaker was a Christian and became a convert to the Roman Catholic Church (1930). In relation to that he was a member of the Pontifical Academy of Sciences from 1936 onward and was president of a Newman Society. Earlier at Cambridge in 1901 he married the daughter of a learned Presbyterian minister. [source: Wikipedia bio]

**Guglielmo Marconi** (1874-1937) Inventor, best known for his development of a radio telegraph system. He sought to use radio waves (discovered by Heinrich Hertz) to create a practical system of "wireless telegraphy"—i.e. the transmission of telegraph messages without connecting wires as used by the electric telegraph. Numerous investigators had been exploring wireless telegraph technologies for over 50 years, but none had proven commercially successful. Nor did Marconi discover any new and revolutionary principle in his wireless-telegraph system, but rather he assembled and improved a number of components, unified and adapted them to his system. In 1895, after increasing the length of the transmitter and receiver antennas, and arranging them vertically, and positioning the antenna so that it touched the ground, he was able to transmit signals over a hill, a distance of approximately 1.5 kilometres (0.93 mi). By March 1897, Marconi had transmitted Morse code signals over a distance of about 6 kilometres (3.7 mi). On 13 May 1897, Marconi sent the first ever wireless communication over open sea: a distance of 6 kilometres (3.7 mi). On 15 November 1899 the *St. Paul* became the first ocean liner to report her imminent arrival by wireless when Marconi's Needles station contacted her sixty-six nautical miles off the English coast. On 17 December 1902, a transmission from the Marconi station in Glace Bay, Nova Scotia, Canada, became the first radio message to cross the Atlantic from North America. Baptized as a Catholic, he was also a member of the Anglican Church, being married into it; however, he still received a Catholic annulment. [source: Wikipedia bio] Marconi stated:

> The more I work with the powers of Nature, the more I feel God's benevolence to man; the closer I am to the great truth that everything is dependent on the Eternal Creator and Sustainer; the more I feel that the so-called science, I am occupied with, is nothing but an expression of the Supreme Will, which aims at bringing people closer to each other in order to help them better understand and improve themselves.

(cited in Maria Cristina Marconi, *Mio Marito Guglielmo*, R.C.S. Libri e Grandi Opere S.p.A. Milano: Rizzoli. [Trans. Raina Castoldi], 1995, 244)

Science alone is unable to explain many things, and most of all, the greatest of mysteries – the mystery of our existence. I believe, not only as a Catholic, but also as a scientist.
(cited in Louis LaRavoire Morrow, "Some Catholic Scientists," in *My Catholic Faith: A Manual of Religion* [Kenosha, Wisconsin, USA: My Mission House: 1949], 14a)

**Sir James Hopwood Jeans** (1877-1946) Physicist, astronomer and mathematician who made important contributions in many areas of physics, including quantum theory, the theory of radiation and stellar evolution. His scientific reputation is grounded in the monographs *The Dynamical Theory of Gases* (1904), *Theoretical Mechanics* (1906), and *Mathematical Theory of Electricity and Magnetism* (1908). After retiring in 1929, he wrote a number of books for the lay public, including *The Stars in Their Courses* (1931), *The Universe Around Us*, *Through Space and Time* (1934), *The New Background of Science* (1933), and *The Mysterious Universe*. These books made Jeans fairly well known as an expositor of the revolutionary scientific discoveries of his day, especially in relativity and physical cosmology. One of Jeans' major discoveries, named Jeans length, is a critical radius of an interstellar cloud in space. Another version of this equation, called Jeans mass or Jeans instability, explains the critical mass a cloud must attain before being able to collapse. He also helped to discover the Rayleigh-Jeans law, which relates the energy density of blackbody radiation to the temperature of the emission source. [source: Wikipedia bio]

**Oswald Theodore Avery** (1877-1955) Physician and medical researcher; one of the first molecular biologists and a pioneer in immunochemistry, but he is best known for his discovery in 1944, with his co-workers Colin MacLeod and Maclyn McCarty, that DNA is the material of which genes and chromosomes are made. For many years, genetic information was thought to be contained in cell protein. Continuing the research done by Frederick Griffith in 1927, Avery worked with MacLeod and McCarty on the mystery of inheritance. Alfred Hershey and Martha Chase furthered Avery's research in 1952 with the Hershey-Chase experiment. These experiments paved the way for Watson and Crick's discovery of the helical structure of DNA, and thus the birth of modern genetics and molecular biology. Nobel laureate Joshua Lederberg stated that Avery and his laboratory provided "the historical platform of modern DNA research" and "betokened the molecular revolution in genetics and biomedical science generally." [source: Wikipedia bio]

**Albert Einstein** (1879-1955) Theoretical physicist and philosopher who is widely regarded as one of the most influential and best-known scientists and intellectuals of all time. He is often regarded as the father of modern physics. He received the 1921 Nobel Prize in Physics "for his services to Theoretical Physics, and especially for his discovery of the law of the photoelectric effect". His many contributions to physics include the special and general theories of relativity, the founding of relativistic cosmology, the first post-Newtonian expansion, the explanation of the perihelion precession of Mercury, the prediction of the deflection of light by gravity (gravitational lensing), the first fluctuation dissipation theorem which explained the Brownian motion of molecules, the photon theory and the wave-particle duality, the quantum theory of atomic motion in solids, the zero-point energy concept, the semi-classical version of the Schrödinger equation, and the quantum theory of a monatomic gas which predicted Bose–Einstein condensation. Einstein published more than 300 scientific and over 150 non-scientific works; he additionally wrote and commentated prolifically on various philosophical and political subjects. [source: Wikipedia bio] For extensive documentation of his pantheist or "panentheistic" religious views (quite opposed to atheism), see chapter 10.

**Otto Hahn** (1879-1968) Chemist and Nobel laureate who pioneered the fields of radioactivity and radiochemistry. He is regarded as "the father of nuclear chemistry" and the "founder of the atomic age". In February 1921, Otto Hahn published the first report on his discovery of uranium Z (later known as $^{234}$Pa), the first example of nuclear isomerism. "a discovery that was not understood at the time but later became highly significant for nuclear physics", as Walther Gerlach remarked. In the early 1920s, Otto Hahn created a new field of work. Using the "emanation method", which he had recently developed, and the "emanation ability", he founded what became known as "Applied Radiochemistry" for the researching of general chemical and physical-chemical questions. On the evidence of a decisive experiment on 17 December 1938 with his pupil and assistant Fritz Strassmann (the celebrated "radium-barium-mesothorium-fractionation"), Otto Hahn concluded that the uranium nucleus had "burst" into atomic nuclei of medium weight. This was the discovery of nuclear fission. In their second publication on nuclear fission (*Die Naturwissenschaften*, 10 February 1939) Otto Hahn and Fritz Strassmann predicted the existence and liberation of additional neutrons during the fission process, which was proofed as chain reaction by Frédéric Joliot and his team in March 1939. At the end of 1999 the German newsmagazine FOCUS published an inquiry of 500 leading natural scientists, engineers and physicians about the most important scientists of the 20th century. In this poll the experimental chemist Otto Hahn - after the theoretical physicists Albert Einstein and Max Planck - was elected third (with 81 points) and thus the most significant empiric researcher of his time. (FOCUS, No. 52, 1999, p. 103-108). [source: Wikipedia bio]

**Lewis Fry Richardson** (1881-1953) Mathematician, physicist, and meteorologist (Quaker in religion) who pioneered modern mathematical techniques of weather forecasting. He is also noted for his pioneering work on fractals and a method for solving a system of linear equations known as modified Richardson iteration. Richardson's interest in meteorology led him to propose a scheme for weather forecasting by solution of differential equations, the method used today. He was also interested in atmospheric turbulence and performed many terrestrial experiments. The Richardson number, a dimensionless parameter in the theory of turbulence is named after him. [source: Wikipedia bio]

**Sir Alexander Fleming** (1881-1955) Biologist and pharmacologist who published many articles on bacteriology, immunology and chemotherapy. His best-known achievements are the discovery of the enzyme lysozyme in 1923 and the antibiotic substance penicillin from the fungus *Penicillium notatum* in 1928. Fleming's accidental discovery and isolation of penicillin marks the start of modern antibiotics. [source: Wikipedia bio]

**Sir Arthur Stanley Eddington** (1882-1944) Astrophysicist. The Eddington limit, the natural limit to the luminosity of stars, or the radiation generated by accretion onto a compact object, is named in his honour. He wrote a number of articles which announced and explained Einstein's theory of general relativity to the English-speaking world and also conducted an expedition to observe the Solar eclipse of 29 May 1919 that provided one of the earliest confirmations of relativity. According to the theory of general relativity, stars with light rays that passed near the Sun would appear to have been slightly shifted because their light had been curved by its gravitational field. This effect is noticeable only during eclipses, since otherwise the Sun's brightness obscures the affected stars. He collected many of his lectures in his *Mathematical Theory of Relativity* in 1923, which Albert Einstein suggested was "the finest presentation of the subject in any language." He was an early advocate of Einstein's General Relativity, investigated the interior of stars through theory, and developed the first true understanding of stellar processes, and defended his method by pointing to the utility of his results, particularly his important mass-luminosity relation. This had the unexpected result of showing that virtually all stars, including giants and dwarfs, behaved as ideal gases. He demonstrated that the interior temperature of stars must be millions of degrees. In

1924, he discovered the mass-luminosity relation for stars. The confirmation of his estimated stellar diameters by Michelson in 1920 proved crucial in convincing astronomers unused to Eddington's intuitive, exploratory style. His theory appeared in mature form in 1926 as *The Internal Constitution of the Stars*, which became an important text for training an entire generation of astrophysicists. Eddington, a Quaker, was willing to discuss the philosophical and religious implications of the new physics and argued for a deeply-rooted philosophical harmony between scientific investigation and religious mysticism, and also that the positivist nature of modern physics (i.e., relativity and quantum physics) provided new room for personal religious experience and free will. He rejected the idea that science could provide proof of religious propositions. [source: Wikipedia bio]

**Robert Goddard** (1882-1945) Physicist and inventor who is credited with creating and building the world's first liquid-fueled rocket, which he successfully launched on March 16, 1926. Goddard and his team launched 34 rockets between 1926 and 1941, achieving altitudes as high as 2.6 km (1.62 miles) and speeds as high as 885 km/h (550 mph). As both theorist and engineer, Goddard's work anticipated many of the developments that made space flight possible. Two of Goddard's 214 patents — one for a multi-stage rocket design (1915), and another for a liquid-fuel rocket design (1915) — are regarded as important milestones toward spaceflight. His 1919 monograph, *A Method of Reaching Extreme Altitudes*, is considered one of the classic texts of 20th century rocket science. Goddard successfully applied three-axis control, gyroscopes and steerable thrust to rockets, all of which allow rockets to be controlled effectively in flight. In the decades around 1910, radio was a new technology, a fertile field for innovation. In 1911, while working at Clark University, Goddard investigated the effects of radio waves on insulators. In order to generate radio-frequency power, in 1915 he invented a vacuum tube that operated like a cathode-ray tube. This was the first use of a vacuum tube to amplify a signal, preceding even Lee de Forest's claim. [source: Wikipedia bio]

Erwin Schrödinger (1887-1961) Theoretical physicist and one of the fathers of quantum mechanics, famed for a number of important contributions to physics. In 1935, after extensive correspondence with personal friend Albert Einstein, he proposed the Schrödinger's cat thought experiment. In January 1926, Schrödinger published in *Annalen der Physik* the paper "Quantization as an Eigenvalue Problem" on wave mechanics and what is now known as the Schrödinger equation. In this paper he gave a "derivation" of the wave equation for time independent systems, and showed that it gave the correct energy eigenvalues for the hydrogen-like atom. This paper has been universally celebrated as one of the most important achievements of the twentieth century, and created a revolution in quantum mechanics, and indeed of all physics and chemistry. A second paper was submitted just four weeks later that solved the quantum harmonic oscillator, the rigid rotor and the diatomic molecule, and gives a new derivation of the Schrödinger equation. A third paper in May showed the equivalence of his approach to that of Heisenberg and gave the treatment of the Stark effect. A fourth paper in this most remarkable series showed how to treat problems in which the system changes with time, as in scattering problems. These papers were the central achievement of his career and were at once recognized as having great significance by the physics community. He wrote about 50 further publications on various topics, including his explorations of unified field theory. One of Schrödinger's lesser-known areas of scientific contribution was his work on color, color perception, and colorimetry (*Farbenmetrik*). In 1920, he published three papers in this area. [source: Wikipedia bio]

> Science is a game – but a game with reality, a game with sharpened knives. If a man cuts a picture carefully into 1000 pieces, you solve the puzzle when you reassemble the pieces into a picture; in the success or failure, both your intelligences compete. In the presentation of a scientific problem, the other player is the good Lord. He has not only set the problem but also has devised the rules of the game – but they are not completely known, half of

them are left for you to discover or to deduce. The uncertainty is how many of the rules God himself has permanently ordained, and how many apparently are caused by your own mental inertia, while the solution generally becomes possible only through freedom from its limitations. This is perhaps the most exciting thing in the game.

(cited in Walter John Moore, *Schrödinger: Life and Thought* [Cambridge: Cambridge University Press: 1990], 348)

The grave error in a technically directed cultural drive is that it sees its highest goal in the possibility of achieving an alteration of Nature. It hopes to set itself in the place of God, so that it may force upon the divine will some petty conventions of its dust-born mind.

(cited in Walter John Moore, *ibid.*, 349)

Consciousness cannot be accounted for in physical terms. For consciousness is absolutely fundamental. It cannot be accounted for in terms of anything else.

(Erwin Schrödinger, "General Scientific and Popular Papers," in *Collected Papers*, Vol. 4. Vienna: Austrian Academy of Sciences. Friedr. Vieweg & Sohn, Braunschweig/Wiesbaden 1984, 334)

**Sir Ronald Aylmer Fisher** (1890-1962) Statistician, evolutionary biologist, and geneticist, described by Anders Hald as "a genius who almost single-handedly created the foundations for modern statistical science." He wrote the ground-breaking treatise, "The Correlation Between Relatives on the Supposition of Mendelian Inheritance" in 1916. This paper laid the foundation for what came to be known as biometrical genetics, and introduced the very important methodology of the analysis of variance, which was a considerable advance over the correlation methods used previously. The paper showed very convincingly that the inheritance of traits measurable by real values, the values of continuous variables, is consistent with Mendelian principles. He pioneered the principles of the design of experiments. His first book, *Statistical Methods for Research Workers* went through many editions and became a standard reference work for scientists in many disciplines. In 1935, this was followed by *The Design of Experiments*, which also became a standard. Fisher invented the technique of maximum likelihood and originated the concepts of sufficiency, ancillarity, Fisher's linear discriminator and Fisher information. His 1924 article "On a distribution yielding the error functions of several well known statistics" presented Karl Pearson's chi-squared and Student's t in the same framework as the Gaussian distribution, and his own "analysis of variance" distribution z (more commonly used today in the form of the F distribution). These contributions easily made him a major figure in 20th century statistics. He initiated the field of non-parametric statistics. His work on the theory of population genetics also made him one of the three great figures of that field, together with Sewall Wright and J.B.S. Haldane, and as such was one of the founders of the neo-Darwinian modern evolutionary synthesis. In addition to founding modern quantitative genetics with his 1918 paper, he was the first to use diffusion equations to attempt to calculate the distribution of gene frequencies among populations. He pioneered the estimation of genetic linkage and gene frequencies by maximum likelihood methods, and wrote early papers on the wave of advance of advantageous genes and on clines of gene frequency. Fisher was the original author of the idea of heterozygote advantage, which was later found to play a frequent role in genetic polymorphism. He famously showed that the probability of a mutation increasing the fitness of an organism decreases proportionately with the magnitude of the mutation. He introduced the concept of Fisher information in 1925, some years before Shannon's notions of information and entropy. He was a member of the Church of England and knew the Bible well. H. Allen Orr describes him as "deeply devout Anglican who, between founding modern statistics and population genetics, penned articles for church magazines" in the *Boston Review*. [source: Wikipedia bio]

**Arthur Holly Compton** (1892-1962) Physicist. Around 1913, Arthur Compton devised a demonstration method for the Earth's rotation. In 1918, he began studying X-ray scattering. In 1922, while on faculty at Washington University in St. Louis, Compton found that X-ray wavelengths increase due to scattering of the radiant energy by "*free electrons*". The scattered quanta have less energy than the quanta of the original ray. This discovery, known as the "Compton effect," or "Compton scattering" demonstrates the "*particle*" concept of electromagnetic radiation and earned Compton the Nobel Prize in physics in 1927. Compton developed the method for observing at the same instant individual scattered X-ray photons and the recoil electrons (developed with Alfred W. Simon). For a time Arthur Compton was a deacon at a Baptist church. [source: Wikipedia bio] Commenting on the first verse of the Bible in *Chicago Daily News* (12 April 1936; magazine section), Arthur Compton stated his religious views:

> For myself, faith begins with the realization that a supreme intelligence brought the universe into being and created man. It is not difficult for me to have this faith, for it is incontrovertible that where there is a plan there is intelligence. An orderly, unfolding universe testifies to the truth of the most majestic statement ever uttered: 'In the beginning God...' [Genesis 1, 1].
> 
> If religion is to be acceptable to science it is important to examine the hypothesis of an Intelligence working in nature. The discussion of the evidences for an intelligent God is as old as philosophy itself. The argument on the basis of design, though trite, has never been adequately refuted. On the contrary, as we learn more about our world, the probability of its having resulted by chance processes becomes more and more remote, so that few indeed are the scientific men of today who will defend an atheistic attitude.

(*The Freedom of Man* [The Terry Lectures] [New Haven: Yale University Press: 1935], 73)

Monsignor Georges Lemaître (1894-1966; priest) Professor of physics and astronomer at the Catholic University of Louvain, who proposed what became known as the Big Bang theory of the origin of the Universe, which he called his "hypothesis of the primeval atom". In 1923 as a graduate student in astronomy at the University of Cambridge, he worked with Arthur Eddington who initiated him into modern cosmology, stellar astronomy, and numerical analysis. In 1927 he published in the *Annals of the Scientific Society of Brussels*, the paper, "A homogeneous Universe of constant mass and growing radius accounting for the radial velocity of extragalactic nebulae". In this report, he presented his new idea of an expanding Universe but not yet that of the primeval atom. Instead, the initial state was taken as Einstein's own finite-size static universe model. Lemaître was a pioneer in applying Albert Einstein's theory of general relativity to cosmology. In the same 1927 article, which preceded Edwin Hubble's landmark article by two years, Lemaître derived what became known as Hubble's law and proposed it as a generic phenomenon in relativistic cosmology. Lemaître also estimated the numerical value of the Hubble constant. However, the data used by Lemaître did not allow him to prove that there was an actual linear relation, which Hubble did two years later. In 1927, Einstein, while not taking exception to the mathematics of Lemaître's theory, refused to accept the idea of an expanding universe; Lemaître recalled him commenting "Your math is correct, but your physics is abominable." But after Hubble's discovery was published, Einstein quickly and publicly endorsed Lemaître's theory, helping both the theory and its proposer get fast recognition. In 1930, Eddington published in the *Monthly Notices of the Royal Astronomical Society* a long commentary on Lemaître's 1927 article, in which he described the latter as a "brilliant solution" to the outstanding problems of cosmology. Lemaître was invited in 1931 to take part in a meeting of the British Association on the relation between the physical Universe and spirituality. There he proposed that the Universe expanded from an initial point, which he called the "Primeval Atom" and developed in a report published in *Nature* in 1931. Lemaître himself also described his theory as "the Cosmic Egg exploding at the moment of the creation"; it became better known as the "Big Bang theory," a term coined by Fred Hoyle. This proposal met skepticism from his fellow scientists at the time. Eddington found Lemaître's notion unpleasant. Einstein found it suspect because he deemed it unjustifiable from a physical point of view. On the other hand, Einstein encouraged Lemaître to look into the possibility of models of non-isotropic expansion, so it's clear he was not altogether dismissive of the concept. He also appreciated Lemaître's argument that a static-Einstein model of the universe could not be sustained indefinitely into the past. In 1933, when he resumed his theory of the expanding Universe and published a more detailed version in the *Annals of the Scientific Society of Brussels*, Lemaître would achieve his greatest glory. On March 17, 1934, Lemaître received the Francqui Prize, the highest Belgian scientific distinction, from King Léopold III. His proposers were Albert Einstein, Charles de la Vallée-Poussin and Alexandre de Hemptinne. The members of the international jury were Eddington, Langevin and Théophile de Donder. Lemaître's theory changed the course of cosmology. This was because Lemaître was well acquainted with the work of astronomers, and designed his theory to have testable implications and to be in accord with observations of the time, in particular, to explain the observed redshift of galaxies and the linear relation beween distances and velocities; he proposed his theory at an opportune time, since Edwin Hubble would soon publish his velocity-distance relation that strongly supported an expanding universe and, consequently, the Big Bang theory; he had studied under Arthur Eddington, who made sure that Lemaître got a hearing in the scientific community. Lemaître was the first to propose that the expansion explains the redshift of galaxies. He further concluded that an initial "creation-like" event must have occurred. In the 1980s, Alan Guth and Andrei Linde modified this theory by adding to it a period of inflation. In 1933, Lemaître found an important inhomogeneous solution of Einstein's field equations describing a spherical dust cloud, the Lemaitre–Tolman metric. He was also a remarkable algebraicist and arithmetical calculator. Since 1930, he used the most powerful calculating machines of the time like the Mercedes. In 1958, he introduced at the University at Burroughs E 101, the University's first electronic computer. Lemaître kept a strong interest in the development of computers and, even more, in the problems of language and programming. He died on June 20, 1966, shortly after having learned of the discovery of cosmic microwave background radiation, which provided further evidence for his intuitions about the birth of the Universe. [source: Wikipedia bio] Despite this high praise, there were some problems with Lemaître's theory. For one, Lemaître's calculated rate of expansion did not work out. If the universe was expanding at a steady rate, the time it had taken to cover its radius was too short to allow for the formation of the stars and planets. Lemaître solved this problem by expropriating Einstein's cosmological constant. Where Einstein had used it in an attempt to keep the universe at a steady size, Lemaître used it to speed up the expansion of the universe over time. After Arthur Eddington died in 1944, Cambridge University became a center of opposition to Lemaître's theory of the Big Bang. In 1964 there was a significant breakthrough that confirmed some of Lemaître's theories. Workers at Bell Laboratories in New Jersey were tinkering with a radio telescope when they discovered a frustrating kind of microwave interference. It was equally strong whether they pointed their telescope at the center of the galaxy or in the opposite direction. What was more, it always had the same wavelength and it always conveyed the same source temperature. This accidental discovery required the passage of several months for its importance to sink in. Eventually, it won Arno Penzias the Nobel Prize in physics. This microwave interference came to be recognized as cosmic background radiation, a remnant of the Big Bang. When word of the 1998 Berkeley discovery that the universe is expanding at an increasing rate first reached Stephen Hawking, he said it was too preliminary to be taken seriously. Later, he changed his mind. "I have now had more time to consider the observations, and they look quite good," he told *Astronomy* magazine (October 1999). "This led me to reconsider my theoretical prejudices." [source: Mark Midbon, "'A Day Without Yesterday': Georges Lemaitre and the Big Bang," *Commonweal* (March 24, 2000): 18-19]

**Hermann Julius Oberth** (1894-1989) Physicist and engineer who, along with Konstantin Tsiolkovsky and Robert H. Goddard, was one of the founding fathers of rocketry and astronautics. In his youthful experiments, he arrived independently at the concept of the multistage rocket. By 1917, he had fired a rocket with liquid propellant. In 1922, Oberth's proposed doctoral dissertation on rocket science was rejected as "utopian". He next had his 92-page work published privately in June 1923 as the somewhat controversial book, *By Rocket into Planetary Space*. By 1929, Oberth had expanded this work to a 429-page book: *Ways to Spaceflight*. In the autumn of 1929, Oberth conducted a static firing of his first liquid-fueled rocket motor, which he named the *Kegeldüse*. In 1953, he published his book *Men in Space*, in which he described his ideas for space-based reflecting telescopes, space stations, electric-powered spaceships, and space suits. [source: Wikipedia bio]

**Edward Arthur Milne** (1896-1950) Astrophysicist and mathematician. Milne's work on the atmospheres of stars extended work done earlier by Schuster in 1905 and by Schwarzschild in 1906. Schuster had studied the transfer of radiation where it was assumed that no absorption was taking place in the atmosphere, while Schwarzschild studied equilibrium states for radiation in an absorbing atmosphere. Milne combined the two approaches and came up with an integral equation of great mathematical interest which is now known as Milne's integral equation. He developed a new form of relativity called kinematic relativity, an alternative to Einstein's general theory of relativity, which also met with considerable opposition. However, his work made people rethink old ideas and led to new approaches to the fundamental concepts of space and time. Milne's books include *Thermodynamics of the Stars* (1930), *Relativity, Gravitation and World-Structure* (1935), and *Kinematic Relativity* (1948). [source: MacTutor bio] He was, as would be the next Rouse Ball Professor of Mathematics (Oxford), Charles Coulson, a pioneer in religion and science discussions, and authored *Modern Cosmology and the Christian Idea of God* (Oxford: Clarendon Press, 1952). [source: Wikipedia bio]

**Werner Heisenberg** (1901-1976) Theoretical physicist who made foundational contributions to quantum mechanics and is best known for asserting the uncertainty principle of quantum theory. In addition, he made important contributions to nuclear physics, quantum field theory, and particle physics. Heisenberg, along with Max Born and Pascual Jordan, set forth the matrix formulation of quantum mechanics in 1925. In early 1929, Heisenberg and Wolfgang Pauli submitted the first of two papers laying the foundation for relativistic quantum field theory. Shortly after the discovery of the neutron by James Chadwick in 1932, Heisenberg submitted the first of three papers on his neutron-proton model of the nucleus. In September 1942, Heisenberg submitted his first paper of a three-part series on the scattering matrix, or S-matrix, in elementary particle physics. The first two papers were published in 1943 and the third in 1944. The S-matrix described only observables, i.e., the states of incident particles in a collision process, the states of those emerging from the collision, and stable bound states; there would be no reference to the intervening states. In the post-war period, Heisenberg continued his interests in cosmic-ray showers with considerations on multiple production of mesons and the unified field theory of elementary particles. From 1957, Heisenberg was interested in plasma physics and the process of nuclear fusion. On 24 March 1973, Heisenberg gave a speech before the Catholic Academy of Bavaria, accepting the Romano Guardini Prize. An English translation of its title is "Scientific and Religious Truth." And its stated goal was "In what follows, then, we shall first of all deal with the unassailability and value of scientific truth, and then with the much wider field of religion, of which--so far as the Christian religion is concerned--Guardini himself has so persuasively written; finally--and this will be the hardest part to formulate--we shall speak of the relationship of the two truths." [source: Wikipedia bio]

The first gulp from the glass of natural sciences will turn you into an atheist, but at the bottom of the glass God is waiting for you. ["Der erste Trunk aus dem Becher der Naturwissenschaft macht atheistisch, aber auf dem Grund des Bechers wartet Gott."]

(cited in Ulrich Hildebrand, "Das Universum - Hinweis auf Gott?", in *Ethos* (die Zeitschrift für die ganze Familie), No. 10, Oktober. Berneck, Schweiz: Schwengeler Verlag AG. Reprinted by permission of the publisher, Schwengeler Verlag AG, 1988, 10)

In the history of science, ever since the famous trial of Galileo, it has repeatedly been claimed that scientific truth cannot be reconciled with the religious interpretation of the world. Although I am now convinced that scientific truth is unassailable in its own field, I have never found it possible to dismiss the content of religious thinking as simply part of an outmoded phase in the consciousness of mankind, a part we shall have to give up from now on. Thus in the course of my life I have repeatedly been compelled to ponder on the relationship of these two regions of thought, for I have never been able to doubt the reality of that to which they point.

(Werner Heisenberg, *Across the Frontiers* [*World Perspectives*, Vol. 48] [New York & San Francisco: Harper and Row Publishers: 1974], 213)

We can console ourselves that the good Lord God would know the position of the particles, and thus He could let the causality principle continue to have validity.

(Heisenberg's last known letter to Einstein, as cited in Gerald Holton, "Werner Heisenberg and Albert Einstein," *Physics Today*, Vol. 53, Issue 7, 2000) [source: excerpt from Tihomir Dimitrov, editor, *50 Nobel Laureates and Other Great Scientists Who Believe in God* (online book, 2008) ]

**John von Neumann** (1903-1957) Mathematician who made major contributions to a vast range of fields, including set theory, functional analysis, quantum mechanics, ergodic theory, continuous geometry, economics and game theory, computer science, numerical analysis, hydrodynamics (of explosions), and statistics, as well as many other mathematical fields. He is generally regarded as one of the greatest mathematicians in modern history. The mathematician Jean Dieudonné called von Neumann "the last of the great mathematicians", while Peter Lax described him as possessing the most "fearsome technical prowess" and "scintillating intellect" of the century. Von Neumann was a pioneer of the application of operator theory to quantum mechanics, in the development of functional analysis, and a key figure in the development of game theory, the concepts of cellular automata, and the universal constructor. Shortly before he died, he invited a Catholic priest, Father Anselm Strittmatter, O.S.B., to visit him for consultation (a move which shocked some of von Neumann's friends). The priest then administered to him the sacrament of anointing. [source: Wikipedia bio]

**Kurt Gödel** (1906-1978) Logician, mathematician, and philosopher: one of the most significant logicians of all time, Gödel made an immense impact upon scientific and philosophical thinking in the 20th century. He is best known for his two incompleteness theorems, published in a paper in 1931 ("On formally undecidable propositions of *Principia Mathematica* and related systems") when he was 25 years of age. The more famous incompleteness theorem states that for any self-consistent recursive axiomatic system powerful enough to describe the arithmetic of the natural numbers (Peano arithmetic), there are true propositions about the naturals that cannot be proved from the axioms. To prove this theorem, Gödel developed a technique now known as Gödel numbering, which codes formal expressions as natural numbers. He also showed that the continuum hypothesis cannot be disproved from the accepted axioms of set theory, if those axioms are consistent. He made important contributions to proof theory by clarifying the connections between classical logic, intuitionistic logic, and modal logic. In 1929, at the age of 23, he completed his doctoral dissertation under Hans Hahn's supervision. In it, Gödel established the completeness of the first-order predicate calculus (this result is known as Gödel's completeness theorem). In 1940, he published his work *Consistency of the axiom of choice and of the generalized continuum-hypothesis with the axioms of set theory* which is a classic of modern mathematics. In that work he introduced the constructible universe, a model of set theory in which the only sets that exist are those that can be constructed from simpler sets. Gödel showed that both the axiom of choice (AC) and the generalized continuum hypothesis (GCH) are true in the constructible universe, and therefore must be consistent with the Zermelo–Fraenkel axioms for set theory (ZF). Gödel and Einstein subsequently developed a strong friendship, and were known to take long walks together to and from the Institute for Advanced Study at Princeton. Economist Oskar Morgenstern recounts that toward the end of his life Einstein confided that his "own work no longer meant much, that he came to the Institute merely...to have the privilege of walking home with Gödel". In 1951, Gödel demonstrated the existence of paradoxical solutions to Albert Einstein's field equations in general relativity. These "rotating universes" would allow time travel and caused Einstein to have doubts about his own theory. His solutions are known as the Gödel metric. In the early 1970s, Gödel circulated among his friends an elaboration of Leibniz's version of Anselm of Canterbury's ontological proof of God's existence. This is now known as Gödel's ontological proof. He was a convinced theist and rejected the notion of others like his friend Albert Einstein that God was impersonal. He believed firmly in an afterlife, stating: "I am convinced of the afterlife, independent of theology. If the world is rationally constructed, there must be an afterlife." In an unmailed answer to a questionnaire, Gödel described his religion as "baptized Lutheran (but not member of any religious congregation). My belief is *theistic*, not pantheistic, following Leibniz rather than Spinoza." [source: Wikipedia bio]

# Chapter Nine

## 115 Scientific Fields of Study Founded or Extraordinarily Advanced by Christian or Theistic Scientists / 34 Prominent Catholic Priest-Scientists and Mathematicians: 1500-1950

[asterisk after dates signifies a Catholic priest]

**Acoustics** Joseph Henry (1797-1878)
Lord Rayleigh (1842-1919)

**Aerodynamics / Aeronautics**
Francesco Lana de Terzi (c. 1631-1687*)
George Cayley (1773-1857)

**Analysis** Leonhard Euler (1707-1783)
Augustin-Louis Cauchy (1789-1857)

Karl Weierstrass (1815-1897)

**Anatomy** Andreas Vesalius (1514-1564)

**Anatomy, Comparative** Georges Cuvier (1769-1832)

**Anesthesiology** James Simpson (1811-1870)

**Antiseptic Surgery** Joseph Lister (1827-1912)

**Applied Science** Gottfried Wilhelm Leibniz (1646-1716)

**Astronautics** Robert Goddard (1882-1945)
Hermann Oberth (1894-1989)

**Astronomy, Big Bang Cosmology** Georges Lemaître (1894-1966*)

**Astronomy, Galactic** William Herschel (1738-1822)

**Astronomy, Heliocentric** Nicolaus Copernicus (1473-1543)

**Astronomy, Physical** Johannes Kepler (1571-1630)

**Astronomy, Solar** Arthur Stanley Eddington (1882-1944)

**Atomic Theory** Roger Boscovich (1711-1787*)
John Dalton (1766-1844)

**Bacteriology** Louis Pasteur (1822-1895)

**Biochemistry** Franciscus Sylvius (1614-1672)
Antoine Lavoisier (1743-1794)

**Biogeography** Alexander von Humboldt (1769-1859)

**Biology / Natural History** John Ray (1627-1705)

**Biology, Molecular** Oswald Avery (1877-1955)
George Wells Beadle (1903-1989)

**Botany** Otto Brunfels (1488–1534)
Carolus Clusius (1526-1609)
Carol Linnaeus (1707-1778)

**Calculus** Blaise Pascal (1623-1662)

**Calculus, Infinitesimal** Pierre de Fermat (c. 1607-1665)
Isaac Newton (1642-1727)
Gottfried Wilhelm Leibniz (1646-1716)

**Cardiology** William Harvey (1578-1657)

**Chemistry** Robert Boyle (1627-1691)

**Chemistry, Agricultural** Joseph Henry Gilbert (1817-1901)

**Chemistry, Isotopic** William Ramsay (1852-1916)

**Chemistry, Nuclear** Otto Hahn (1879-1968)

**Chemistry, Organic** Thomas Anderson (1819-1874)

**Chemistry, Physical** Josiah Willard Gibbs (1839-1903)

**Computer Science** Charles Babbage (1792-1871)
George Boole (1815-1864)

**Cryology** Lord Kelvin (1824-1907)

**Cytology** Robert Hooke (1635-1703)
Jean Baptiste Carnoy (1836–1899)

**Dimensional Analysis** Lord Rayleigh (1842-1919)

**Dynamics** Isaac Newton (1642-1727)

**Ecology** Carol Linnaeus (1707-1778)

**Electrical Engineering** William Gilbert (1544-1603)
Michael Faraday (1791-1867)
Thomas Edison (1847-1931)

**Electrochemistry** Alessandro Volta (1745-1827)
Humphrey Davy (1778-1829)
Michael Faraday (1791-1867)
**Electrodynamics** André-Marie Ampère (1775-1836)
James Clerk Maxwell (1831-1879)

**Electromagnetics** André-Marie Ampère (1775-1836)
Michael Faraday (1791-1867)
Joseph Henry (1797-1878)
James Clerk Maxwell (1831-1879)

**Electronics** Michael Faraday (1791-1867)
John Ambrose Fleming (1849-1945)

**Electrophysiology** John Eccles (1903-1997)

**Embryology** Julius Caesar Aranzi (1529-1589)
William Harvey (1578-1657)
Marcello Malpighi (1628-1694)

**Energetics** Lord Kelvin (1824-1907)
Josiah Willard Gibbs (1839-1903)

**Entomology** William Kirby (1759-1850)

**Entomology, of Living Insects** Henri Fabre (1823-1915)

**Epidemiology** Girolamo Fracastoro (1478-1553)

**Evolution / Natural Selection** Charles Darwin (1809-1882)

**Field Theory** Michael Faraday (1791-1867)

**Fluid Mechanics** George Stokes (1819-1903)

**Gas Dynamics** Robert Boyle (1627-1691)

**Genetics** Gregor Mendel (1822-1884*)

**Genetics, Clinical Medical** Victor A. McKusick (1921-2008)

**Genetics, Population** Ronald Fisher (1890-1962)

**Geology** Blessed Nicolas Steno (1638-1686*)
James Hutton (1726-1797)

**Geometry, Analytical** René Descartes (1596-1650)
Pierre de Fermat (c. 1607-1665)

**Geometry, Differential** Carl Friedrich Gauss (1777-1855)

**Geometry, Non-Euclidean** Bernhard Riemann (1826-1866)

**Geophysics** Jose de Acosta (1540-1600*)

**Glaciology** Louis Agassiz (1807-1873)
Arnold Henry Guyot (1807-1884)

**Gynecology** James Simpson (1811-1870)

**Histology** Marcello Malpighi (1628-1694)
Marie François Xavier Bichat (1771-1802)

**Hydraulics** Leonardo da Vinci (1452-1519)
Blaise Pascal (1623-1662)

**Hydrodynamics** Blaise Pascal (1623-1662)

**Hydrography** Matthew Maury (1806-1873)

**Hydrology** Edme Mariotte (c. 1620-1684*)

**Hydrostatics** Galileo Galilei (1564-1642)
Blaise Pascal (1623-1662)

**Ichthyology** Louis Aggasiz (1807-1873)

**Immunology** Edward Anthony Jenner (1749-1823)
Louis Pasteur (1822-1895)
**Laser Science** Charles Hard Townes (b. 1915)
Arthur Schawlow (1921-1999)

**Mathematical Analysis** Leonhard Euler (1707-1783)

**Mechanics, Celestial** Johannes Kepler (1571-1630)

**Mechanics, Classical** Isaac Newton (1642-1727)

**Mechanics, Quantum** Max Planck (1858-1947)
Werner Heisenberg (1901-1976)

**Mechanics, Wave** Erwin Schrödinger (1887-1961)

**Medicine, Modern** William Harvey (1578-1657)

**Meteorology** Evangelista Torricelli (1608-1647)
Lazzaro Spallanzani (1729-1799*)
Matthew Maury (1806-1873)

**Microbiology** Athanasius Kircher (1602–1680*)
Marcello Malpighi (1628-1694)
Antonie Philips van Leeuwenhoek (1632-1723)

**Mineralogy** Georgius Agricola (1494-1555)

**Mineralogy, Optical** David Brewster (1781-1868)

**Model Analysis** Lord Rayleigh (1842-1919)

**Morphology** Johann Wolfgang von Goethe (1749-1832)

**Nanotechnology** Richard Smalley (1943-2005)

**Neurology** Charles Bell (1774-1842)

**Number Theory** Leonhard Euler (1707-1783)
Carl Friedrich Gauss (1777-1855)
**Obstetrics** William Smellie (1697-1763)

**Oceanography** Matthew Maury (1806-1873)

**Optics** Galileo Galilei (1564-1642)
Johannes Kepler (1571-1630)
Francesco Maria Grimaldi (1618-1663*)
James Gregory (1638-1675)
Isaac Newton (1642-1727)

**Ornithology** John Ray (1627-1705)

**Paleontology** John Woodward (1665-1728)

**Paleontology, Vertebrate** Georges Cuvier (1769-1832)

**Pathology** Marie François Xavier Bichat (1771-1802)
Thomas Hodgkin (1798-1866)
Rudolph Virchow (1821-1902)

**Physics, Atomic** Joseph J. Thomson (1856-1940)

**Physics, Classical** Isaac Newton (1642-1727)

**Physics, Experimental** Galileo Galilei (1564-1642)

**Physics, Mathematical** Johannes Kepler (1571-1630)
Christiaan Huygens (1629-1695)

Isaac Newton (1642-1727)

**Physics, Nuclear** Ernest Rutherford (1871-1937)

**Physics, Particle** John Dalton (1766-1844)

**Physiology** William Harvey (1578-1657)

**Probability Theory** Pierre de Fermat (c. 1607-1665)
Blaise Pascal (1623-1662)
Christiaan Huygens (1629-1695)

**Scientific Method** Francis Bacon (1561-1626)
Galileo Galilei (1564-1642)
Pierre Gassendi (1592-1655*)

**Seismology** John Michell (1724-1793)

**Stellar Spectroscopy** Pietro Angelo Secchi (1818-1878*)
Sir William Huggins (1824-1910)

**Stratigraphy** Blessed Nicolas Steno (1638-1686*)

**Surgery** Ambroise Paré (c. 1510-1590)

**Taxonomy** Carol Linnaeus (1707-1778)

**Thermochemistry** Antoine Lavoisier (1743-1794)

**Thermodynamics** James Joule (1818-1889)
Lord Kelvin (1824-1907)

**Thermodynamics, Chemical** Josiah Willard Gibbs (1839-1903)

**Thermodynamics, Statistical** James Clerk Maxwell (1831-1879)

**Thermokinetics** Humphrey Davy (1778-1829)

**Transfinite Mathematics** Bernard Bolzano (1781-1848*)
Augustin-Louis Cauchy (1789-1857)
Karl Weierstrass (1815-1897)
Georg Cantor (1845-1918)

**Transplantology** Alexis Carrel (1873-1944)
Joseph Murray (b. 1919)

**Volcanology** Athanasius Kircher (1602–1680*)
Lazzaro Spallanzani (1729-1799*)
James Dwight Dana (1813-1895)

**Zoology** Conrad Gessner (1516-1565)

\* \* \* \* \*

Francesco Maurolico (1494-1575; Benedictine abbot) Mathematician, astronomer.
Christopher Clavius (1538-1612; Jesuit priest) Mathematician, astronomer.
Jose de Acosta (1540-1600; Jesuit priest) Geophysicist, meteorologist.
Christopher Scheiner (1575-1650; Jesuit priest) Astronomer, optician.
Benedetto Castelli (1577-1644; Benedictine abbot) Mathematician; hydraulics.
Nicolas Zucchi (1586-1670; Jesuit priest) Optician.
Johann Baptist Cysat (c. 1587-1657; Jesuit priest) Astronomer (expert on comets).
Giovanni Battista Zupi (c. 1590-1650; Jesuit priest) Astronomer.
Pierre Gassendi (1592-1655; priest) Astronomer.
Giovanni Battista Riccioli (1598-1671; Jesuit priest) Physicist.
Jacques de Billy (1602-1679; Jesuit priest) Mathematician, critic of astrology.
Athanasius Kircher (1602–1680; Jesuit priest) Geologist, microbiologist, inventor, Egyptologist, medical theorist (infectious disease).

Juan Caramuel y Lobkowitz (1606-1682; Cistercian and archbishop) Mathematics, astronomy, physics, and probability theory.
André Tacquet (1612-1660; Jesuit priest) Mathematician.
Francesco Maria Grimaldi (1618-1663; Jesuit priest) Optician.
Jean-Felix Picard (1620-1682; Jesuit priest) Geologist (size of the earth), astronomer, inventor.
Edme Mariotte (c. 1620-1684; priest) Physicist, chemist, optician, hydrology.
Francesco Lana de Terzi (c. 1631-1687; Jesuit priest) Aeronautics.
Blessed Nicolas Steno (1638-1686; bishop) Geology (particularly stratigraphy), mineralogist.
Pierre Varignon (1654-1722; Jesuit priest) Statics and mechanics.
Giovanni Girolamo Saccheri (1667-1733; Jesuit priest) Mathematician.
Jean-Antoine Nollet (1700-1770; priest) Electricity, osmosis.
Vincent Riccati (1707-1775; Jesuit priest) Mathematician.
Roger Joseph Boscovich (1711-1787; Jesuit priest) Physicist (atomic theory), astronomer, field theory.
Christian Mayer (1719-1783; Jesuit priest) Astronomer.
Lazzaro Spallanzani (1729-1799; priest) Microbiologist, volcanologist, meteorologist, biologist.
Juan Molina (1740-1829; Jesuit priest) Biochemist; anticipator of evolution.
Giuseppe Piazzi (1746-1826; priest) Astronomer.
Bernard Bolzano (1781-1848; priest) Mathematician and philosopher of science.
Pietro Angelo Secchi (1818-1878; Jesuit priest) Astronomer (especially spectroscopy and the sun).
Gregor Johann Mendel (1822-1884; Augustinian priest) Father of genetics.
Blessed Francesco Faà di Bruno (1825-1888) Mathematician.
Armand David (1826-1900; Lazarist missionary priest) Zoologist and a botanist in China.
Monsignor Georges Lemaître (1894-1966; priest) Astronomer and mathematician; first developed the Big Bang Theory in cosmology.

# Chapter Ten

## Albert Einstein's "Cosmic Religion"

Philosophically, God's existence is something that is reasoned to (as with all other propositions whatever, as well). In a larger epistemology, including religious faith, it is not. I would argue that man is inherently religious (anthropology easily bears this out), so that the religious impulse must be stifled in an atheist. It is already there.

If even rigorous philosophical and scientific minds like David Hume and Einstein look at the universe and immediately sees some sort of Intelligence behind it (though not the *Christian* God), surely there is something to even Paul's assertion of the "plainness" of God's existence, in Romans 1. Hume even stated that "no rational enquirer can, after serious reflection, suspend his belief a moment with regard to the primary principles of genuine Theism and Religion . . ." Einstein made a number of such statements:

> My comprehension of God comes from the deeply felt conviction of a superior intelligence that reveals itself in the knowable world. In common terms, one can describe it as 'pantheistic' (Spinoza).
>
> (Answer to the question, "What is your understanding of God?" *Kaizo*, 5, no. 2, 1923, 197; in Alice Calaprice, editor, <u>The Expanded Quotable Einstein</u>, Princeton University Press, 2000, 203)

Now, I would ask an atheist: whence comes Einstein's "deeply felt conviction"? Is it a philosophical reason or the end result of a syllogism? He simply *has* it. It is an intuitive or instinctive feeling or "knowledge" or "sense of wonder at the incredible, mind-boggling marvels of the universe". Atheists don't possess this intuition, but my point is that it is not utterly implausible or unable to be held by even the most rigorous, "non-dogmatic" intellects, such as Einstein and Hume. And the atheist has to *account* for that fact somehow, it seems to me.

> My religiosity consists of a humble admiration of the infinitely superior spirit that reveals itself in the little that we can comprehend about the knowable world. That deeply emotional conviction of the presence of a superior reasoning power, which is revealed in the incomprehensible universe, forms my idea of God.
>
> (Calaprice, *ibid.*, 204 / To a banker in Colorado, 1927. *Einstein Archive* 48-380; also quoted in Helen Dukas and Banesh Hoffmann, <u>Albert Einstein, the Human Side</u> [Princeton Univ. Press, 1981], 66, and in the *New York Times* obituary, April 19, 1955)
>
> I believe in Spinoza's God who reveals himself in the harmony of all that exists, but not in a God who concerns himself with the fate and actions of human beings.

> (*Ibid.*, 204 / Telegram to a Jewish newspaper, 1929; to Rabbi Herbert S. Goldstein of the Institutional Synagogue in New York . *Einstein Archive* 33-272)

What do atheists think Einstein meant here when he used the word "believe"? Do they think he had an elaborate argument that ended in his conclusion: "I believe in Spinoza's God"?

> I am of the opinion that all the finer speculations in the realm of science spring from a deep religious feeling.
>
> (Calaprice, *ibid.*, 206 / *Forum and Century* 83, 1930, 373)

What does Einstein mean by "deep religious feeling"? Is this a philosophical and/or demonstrable or provable concept? Or is it more like an intuition? How can it be epistemically justified? How can a man like Einstein *hold* such a view in the first place, according to the atheist? Perhaps he himself provides an answer of sorts:

> It is very difficult to elucidate this [cosmic religious] feeling to anyone who is entirely without it . . . In my view, it is the most important function of art and science to awaken this feeling and keep it alive in those who are receptive to it.
>
> (Calaprice, *ibid.*, 207 / <u>*Cosmic Religion*</u>, 1931, 48-49)

In what way would an atheist think Einstein would say such people are "deficient"? He denies that a personal God put this knowledge in people, yet on the other hand he clearly assumes it is innate, normal, and self-evident. How can he do that?

> [T]he belief in the existence of basic all-embracing laws in Nature also rests on a sort of faith. All the same this faith has been largely justified so far by the success of scientific research. But, on the other hand, everyone who

is seriously involved in the pursuit of science becomes convinced that a spirit is manifest in the laws of the universe -- a spirit vastly superior to that of man, and one in the face of which we with our modest powers must feel humble. In this way the pursuit of science leads to a religious feeling of a special sort, which is indeed quite different from the religiosity of someone more naive.

(To student Phyllis Right, who asked if scientists pray, January 24, 1936. *Einstein Archive* 42-601, 52-337; from Dukas and Hoffman, *ibid.*, pp. 32-33)

At first, then, instead of asking what religion is I should prefer to ask what characterizes the aspirations of a person who gives me the impression of being religious: a person who is religiously enlightened appears to me to be one who has, to the best of his ability, liberated himself from the fetters of his selfish desires and is preoccupied with thoughts, feelings and aspirations to which he clings because of their super-personal value . . . Accordingly a religious person is devout in the sense that he has no doubt of the significance of those super-personal objects and goals which neither require nor are capable of rational foundation . . . If one conceives of religion and science according to these definitions then a conflict between them appears impossible. For science can only ascertain what is, but not what should be, and outside of its domain value judgments of all kinds remain necessary. . . . Now, even though the realms of religion and science in themselves are clearly marked off from each other, nevertheless there exist between the two strong reciprocal relationships and dependencies . . . science can only be created by those who are thoroughly imbued with the aspiration toward truth and understanding. This source of feeling, however, springs from the sphere of religion. To this there also belongs the faith in the possibility that the regulations valid for the world of existence are rational, that is, comprehensible to reason. I cannot conceive of a genuine scientist without that profound faith. The situation may be expressed by an image: science without religion is lame, religion without science is blind. . . . a legitimate conflict between science and religion cannot exist. . . . But whoever has undergone the intense experience of successful advances made in this domain is moved by profound reverence for the rationality made manifest in existence. By way of the understanding he achieves a far-reaching emancipation from the shackles of personal hopes and desires, and thereby attains that humble attitude of mind toward the grandeur of reason incarnate in existence, and which, in its profoundest depths, is inaccessible to man. This attitude, however, appears to me to be religious, in the highest sense of the word.

("Science and Religion": Address at the Conference on Science, Philosophy, and Religion in Their Relation to the Democratic Way of Life, New York, 1940; in *Ideas and Opinions* [Crown: New York, 1954, 1982], p. 46; also published in *Nature*, 146: 605-607 [1940] )

In view of such harmony in the cosmos which I, with my limited human mind, am able to recognize, there are yet people who say there is no God. But what makes me really angry is that they quote me for support of such views.

(*Ibid.*, 214 / reply to German anti-Nazi diplomat and author Hubertus zu Lowenstein around 1941. Quoted in the latter's book, *Towards the Further Shore*, London, 1968, 156)

Then there are the fanatical atheists whose intolerance is the same as that of the religious fanatics, and it springs from the same source . . . They are creatures who can't hear the music of the spheres.

(*Ibid.*, 214 / 7 August 1941. *Einstein Archive* 54-297)

I have found no better expression than 'religious' for confidence in the rational nature of reality, insofar as it is accessible to human reason. Whenever this feeling is absent, science degenerates into uninspired empiricism.

(*Ibid.*, 216 / To Maurice Solovine, 1 January 1951. *Einstein Archive* 21-474; published in *Letters to Solovine*, 119)

The bigotry of the nonbeliever is for me nearly as funny as the bigotry of the believer.

(quoted in Robert N. Goldman, *Einstein's God: Albert Einstein's Quest as a Scientist and as a Jew to Replace a Forsaken God* [Jason Aronson: 1997] )

Many similar utterances of Einstein can be found:

I do not believe in a personal God and I have never denied this but have expressed it clearly. If something is in me which can be called religious then it is the unbounded admiration for the structure of the world so far as our science can reveal it.

(Letter to an atheist [24 March 1954] as quoted in *Albert Einstein: The Human Side* [Princeton University Press: 1981], edited by Helen Dukas and Banesh Hoffman, p. 43)

I'm not an atheist and I don't think I can call myself a pantheist. We are in the position of a little child entering a huge library filled with books in many different languages. The child knows someone must have written those books. It does not know how. The child dimly suspects a mysterious order in the arrangement of the books but doesn't know what it is. That, it seems to me, is the attitude of even the most intelligent human being toward God. We see a universe marvelously arranged and obeying certain laws, but only dimly understand these laws. Our limited minds cannot grasp the mysterious force that moves the constellations.

(From an interview, quoted in *Glimpses of the Great* by G. S. Viereck [Macauley, New York, 1930], cited in Max Jammer, *Einstein and Religion: Physics and Theology* [Princeton University Press, 1999], p. 48)

I have repeatedly said that in my opinion the idea of a personal God is a childlike one. You may call me an agnostic, but I do not share the crusading spirit of the professional atheist whose fervor is mostly due to a painful act of liberation from the fetters of religious indoctrination received in youth. I prefer an attitude of humility corresponding to the weakness of our intellectual understanding of nature and of our own being.

(Letter to Guy H. Raner Jr., 28 September 1949, quoted by Michael R. Gilmore in *Skeptic*, Vol. 5, No. 2)

Try and penetrate with our limited means the secrets of nature and you will find that, behind all the discernible concatenations, there remains something subtle, intangible and inexplicable. Veneration for this force beyond anything that we can comprehend is my religion. To that extent I am, in point of fact, religious.

(Response to atheist, Alfred Kerr [Winter 1927] who after deriding ideas of God and religion at a dinner party in the home of the publisher Samuel Fischer, had queried him "I hear that *you* are supposed to be deeply religious" -- as quoted in *Diaries of a Cosmopolitan: Count Harry Kessler, 1918-1937*, by H. G. Kessler, [Littlehampton Book Services Ltd, 1971 edition] )

The fairest thing we can experience is the mysterious. It is the fundamental emotion which stands at the cradle of true art and true science. He who knows it not and can no longer wonder, no longer feel amazement, is as good as dead, a snuffed-out candle. It was the experience of mystery -- even if mixed with fear -- that engendered religion. A knowledge of the existence of something we cannot penetrate, of the manifestations of the profoundest reason and the most radiant beauty, which are only accessible to our reason in their most elementary forms -- it is this knowledge and this emotion that constitute the truly religious attitude; in this sense, and this sense alone, I am a deeply religious man. . . . Enough for me the mystery of the eternity of life, and the inkling of the marvellous structure of reality, together with the single-hearted endeavour to comprehend a portion, be it ever so tiny, of the reason that manifests itself in nature.

(in *The World As I See It* [1949], reprinted in 2007 [Filiquarian Publishing], pp. 14-15; originally from *What I Believe*, 1930; different translation cited in Jammer, *ibid.*, p. 73)

[C]osmic religious feeling is the strongest and noblest incitement to scientific research. Only those who realize the immense efforts and, above all, the devotion which pioneer work in theoretical science demands, can grasp the strength of the emotion out of which alone such work, remote as it is from the immediate realities of life, can issue. What a deep conviction of the rationality of the universe, and what a yearning to understand, were it but a feeble reflection of the mind revealed in this world, Kepler [Lutheran] and Newton [Arian theist] must have had to enable them to spend years of solitary labour in disentangling the principles of celestial mechanics! . . . Only one who has devoted his life to similar ends can have a vivid realization of what has inspired these men and given them the strength to remain true to their purpose in spite of countless failures. It is cosmic religious feeling that gives a man strength of this sort. A contemporary has said, not unjustly, that in this materialistic age of ours the serious scientific workers are the only profoundly religious people.

You will hardly find one among the profounder sort of scientific minds without a peculiar religious feeling of his own.

(*Ibid.*, p. 37; from his essay, "Religion and Science," *New York Times Magazine*, Fall 1930, section 5, pages 1-2) The men who have laid the foundations of physics on which I have been able to construct my theory are Galileo, Newton, Maxwell, and Lorentz.

(Interview with *The New York Times*, 2 April 1921; cited in Max Jammer, *Einstein and Religion: Physics and Theology* [Princeton University Press, 1999], p. 35)

Speaking of the spirit that informs modern scientific investigations, I am of the opinion that all the finer speculations in the realm of science spring from a deep religious feeling, and that without such feeling they would not be fruitful.

(conversation with J. Murray, early in 1930 in Berlin, in Jammer, *ibid.*, pp. 68-69)

I am enough of an artist to draw freely upon my imagination. Imagination is more important than knowledge. Knowledge is limited. Imagination encircles the world.

("What Life Means to Einstein": Interview with George Sylvester Viereck, *The Saturday Evening Post* [26 October 1929, p. 17])

As a child I received instruction both in the Bible and in the Talmud. I am a Jew, but I am enthralled by the luminous figure of the Nazarene. . . . Jesus is too colossal for the pen of phrasemongers, however artful. No man can dispose of Christianity with a *bon mot*. . . . No one can read the Gospels without feeling the actual presence of Jesus. His personality pulsates in every word. No myth is filled with such life. How different, for instance, is the impression which we receive from an account of legendary heroes of antiquity like Theseus. Theseus and other heroes of his type lack the authentic vitality of Jesus. . . . No man can deny the fact that Jesus existed, nor that his sayings are beautiful. Even if some them have been said before, no one has expressed them so divinely as he.

(Interview with George Sylvester Viereck, 26 October 1929; see also Denis Brian, *Einstein — A Life* [John Wiley & Sons, Inc., New York, 1996], pp. 277-278)

What separates me from most so-called atheists is a feeling of utter humility toward the unattainable secrets of the harmony of the cosmos.

("Einstein and Faith," *Time Magazine*, 5 April 2007)

The fanatical atheists are like slaves who are still feeling the weight of their chains which they have thrown off after hard struggle. They are creatures who--in their grudge against traditional religion as the "opium of the masses"-- cannot hear the music of the spheres.

("Einstein and Faith," *Time Magazine*, 5 April 2007)

It is true that convictions can best be supported with experience and clear thinking. On this point one must agree unreservedly with the extreme rationalist. The weak point of his conception is, however, this, that those convictions which are necessary and determinant for our conduct and judgments cannot be found solely along this solid scientific way.

For the scientific method can teach us nothing else beyond how facts are related to, and conditioned by, each other. The aspiration toward such objective knowledge belongs to the highest of which man is capable, and you will certainly not suspect me of wishing to belittle the achievements and the heroic efforts of man in this sphere. Yet it is equally clear that knowledge of what is does not open the door directly to what should be. One can have the clearest and most complete knowledge of what is, and yet not be able to deduct from that what should be the goal of our human aspirations. Objective knowledge provides us with powerful instruments for the achievements of certain ends, but the ultimate goal itself and the longing to reach it must come from another source. And it is hardly necessary to argue for the view that our existence and our activity acquire meaning only by the setting up of such a goal and of corresponding values. The knowledge of truth as such is wonderful, but it is so little capable of acting as a guide that it cannot prove even the justification and the value of the aspiration toward that very knowledge of truth. Here we face, therefore, the limits of the purely rational conception of our existence.. . . .

The highest principles for our aspirations and judgments are given to us in the Jewish-Christian religious tradition. It is a very high goal which, with our weak powers, we can reach only very inadequately, but which gives a sure foundation to our aspirations and valuations. . . .

("Science and Religion," cited in Einstein's *Ideas and Opinions*, pp. 41-49; from an address at Princeton Theological Seminary, 19 May 1939. It was also published in *Out of My Later Years* [New York: Philosophical Library, 1950].

Does there truly exist an insuperable contradiction between religion and science? Can religion be superseded by science? The answers to these questions have, for centuries, given rise to considerable dispute and, indeed, bitter fighting. Yet, in my own mind there can be no doubt that in both cases a dispassionate consideration can only lead to a negative answer.

. . . the function of setting up goals and passing statements of value transcends its domain. While it is true that science, to the extent of its grasp of causative connections, may reach important conclusions as to the compatibility and incompatibility of goals and evaluations, the independent and fundamental definitions regarding goals and values remain beyond science's reach.. . .

There are many such questions which, from a rational vantage point, cannot easily be answered or cannot be answered at all. Yet, I do not think that the so-called "relativistic" viewpoint is correct, not even when dealing with the more subtle moral decisions.. . .

The interpretation of religion, as here advanced, implies a dependence of science on the religious attitude, a relation which, in our predominantly materialistic age, is only too easily overlooked. While it is true that scientific results are entirely independent from religious or moral considerations, those individuals to whom we owe the great creative achievements of science were all of them imbued with the truly religious conviction that this universe of ours is something perfect and susceptible to the rational striving for knowledge. If this conviction had not been a strongly emotional one and if those searching for knowledge had not been inspired by Spinoza's *Amor Dei Intellectualis*, they would hardly have been capable of that untiring devotion which alone enables man to attain his greatest achievements.

("Religion and Science: Irreconcilable?": response to a greeting sent by the Liberal Ministers' Club of New York City. Published in *The Christian Register*, June, 1948.

Published in *Ideas and Opinions* [Crown Publishers, Inc., New York, 1954])

GUEST: I have a letter that Albert Einstein wrote to my father in 1943. In 1940, my father read a "Time Magazine" article that stated that Einstein was quoted as saying that the only social institution that stood up to Nazism was the Christian Church. My father is a Presbyterian minister in a little northern Michigan town called Harbor Springs. And he quoted Einstein in a sermon, and a member of the congregation wrote my father a letter saying, "Where did you get your information?" So my father wrote "Time Magazine" and "Time Magazine" wrote him back, and I have that letter, too, but they didn't give the source, so my father wrote Einstein and he wrote back, saying, yes, he did say that the Christian Church was standing up to Hitler and Nazism.

[ . . . ]

APPRAISER: The second reason I really like this story is that your dad kept all the supporting material behind the letter that he eventually got from Einstein confirming, "Yes, I did say this about the Christian Church. It is the only social institution that could stand up to the Nazi regime." . . . If you had brought this letter in without the supporting documents, I would have looked at it, and it says, "It's true that I made a statement which corresponds approximately with the text you quoted. I made this statement during the first years of the Nazi regime-- much earlier than 1940-- and my expressions were a little more moderate." And I would say, "Well, that's a nice typed letter from Einstein, says something about Nazis," but I wouldn't really know what he was talking about if your father had not saved all the material that is appropriate to it.

("1943 Albert Einstein Letter," *Antiques Roadshow* [PBS], 19 May 2008; the letter was appraised at $5000)

Our time is distinguished by wonderful achievements in the fields of scientific understanding and the technical application of those insights. Who would not be cheered by this? But let us not forget that human knowledge and skills alone cannot lead humanity to a happy and dignified life. Humanity has every reason to place the proclaimers of high moral standards and values above the discoverers of objective truth. What humanity owes to personalities like Buddha, Moses, and Jesus ranks for me higher than all the achievements of the enquiring and constructive mind.
    What these blessed men have given us we must guard and try to keep alive with all our strength if humanity is not to lose its dignity, the security of its existence, and its joy in living.

(From a written statement [September 1937] as quoted in Helen Dukas and Banesh Hoffman, editors, *Albert Einstein: The Human Side* [Princeton University Press: 1981] )

All religions, arts and sciences are branches of the same tree. All these aspirations are directed toward ennobling man's life, lifting it from the sphere of mere physical existence and leading the individual towards freedom. It is no mere chance that our older universities developed from clerical schools. Both churches and universities — insofar as they live up to their true function — serve the ennoblement of the individual. They seek to fulfill this great task by spreading moral and cultural understanding, renouncing the use of brute force.

(In "Moral Decay" [1937], also published in *Out of My Later Years* [1950] )

The longing to behold this pre-established harmony is the source of the inexhaustible patience and perseverance with which Planck has devoted himself, as we see, to the most general problems of our science, refusing to let himself be diverted to more grateful and more easily attained ends. I have often heard colleagues try to attribute this attitude of his to extraordinary will-power and discipline -- wrongly, in my opinion. The state of mind which enables a man to do work of this kind is akin to that of the religious worshiper or the lover; the daily effort comes from no deliberate intention or program, but straight from the heart.

("Principles of Research": address by Albert Einstein in 1918 for the Physical Society, Berlin, on the occasion of Max Planck's sixtieth birthday)

One may say "the eternal mystery of the world is its comprehensibility."

("Physics and Reality" in *Journal of the Franklin Institute* [March 1936]; reprinted in *Out of My Later Years* [1956] )

I fully agree with you about the significance and educational value of methodology as well as history and philosophy of science. So many people today — and even professional scientists — seem to me like someone who has seen thousands of trees but has never seen a forest. A knowledge of the historic and philosophical background gives that kind of independence from prejudices of his generation from which most scientists are suffering. This independence created by philosophical insight is — in my opinion — the mark of distinction between a mere artisan or specialist and a real seeker after truth.

(Letter to Robert A. Thorton, Physics Professor at University of Puerto Rico: 7 December 1944; EA-674, *Einstein Archive*, Hebrew University, Jerusalem)

What I am really interested in is knowing whether God could have created the world in a different way; in other words, whether the requirement of logical simplicity admits a margin of freedom.

(in Jammer, *ibid.*, p. 124)


I want to know how God created this world. I'm not interested in this or that phenomenon, in the spectrum of this or that element. I want to know His thoughts, the rest are details.

(E. Salaman, "A Talk with Einstein," *The Listener* 54 [1955]: 370-371)

I have never imputed to Nature a purpose or a goal, or anything that could be understood as anthropomorphic. What I see in Nature is a magnificent structure that we can comprehend only very imperfectly, and that must fill a thinking person with a feeling of "humility." This is a genuinely religious feeling that has nothing to do with mysticism.

(Reply to a letter: 1954 or 1955; from Helen Dukas and Banesh Hoffmann, *Albert Einstein, the Human Side* [Princeton Univ. Press, 1981], p. 39)

You find it strange that I consider the comprehensibility of the world (to the extent that we are authorized to speak of such a comprehensibility) as a miracle or an eternal mystery. Well a priori one should expect a chaotic world which cannot be grasped by the mind in anyway. One could (yes one should) expect the world to be subjected to law only to the extent that we order it through our intelligence. Ordering of this kind would be like the alphabetical ordering of the words of a language. By contrast, the kind of order created by Newton's theory of gravitation, for instance, is wholly different. Even if the axioms of the theory are proposed by man, the success of such a project presupposes a high degree of ordering of the objective world, and this could not be expected a priori. That is the "miracle" which is being constantly re-enforced as our knowledge expands.

There lies the weaknesss of positivists and professional atheists who are elated because they feel that they have not only successfully rid the world of gods but "bared the miracles." Oddly enough, we must be satisfied to acknowledge the "miracle" without there being any legitimate way for us to approach it.

(Letter to Maurice Solovine; from Robert N. Goldman, *Einstein's God—Albert Einstein's Quest as a Scientist and as a Jew to Replace a Forsaken God* [Joyce Aronson Inc.; Northvale, New Jersey; 1997], p. 24)

The idea of a personal God is quite alien to me and seems even naive. However, I am also not a "Freethinker" in the usual sense of the word because I find that this is in the main an attitude nourished exclusively by an opposition against naive superstition. My feeling is insofar religious as I am imbued with the consciousness of the insufficiency of the human mind to understand deeply the harmony of the Universe which we try to formulate as "laws of nature." It is this consciousness and humility I miss in the Freethinker mentality. Sincerely yours, Albert Einstein.

(Letter to A. Chapple, Australia, 23 February 1954; *Einstein Archive* 59-405; also quoted in Otto Nathan and Heinz Norden, *Einstein on Peace*, [Random House Value Publishing; Avenel 1981 edition], p. 510)

## Chapter Eleven

## The Galileo Case: Historical Facts and Neglected Considerations vs. Secular Revisionist Myths

1) The censure of the astronomer Galileo (1564-1642) in 1616 and 1633 may be the most notorious and famous Catholic error ever made, and the favorite (myth-filled) tale of those who believe religion and science are inexorably opposed. Catholic dogma had never enshrined geocentrism, and Galileo (a faithful Catholic) had been supported by many notable churchmen, including three popes. Indeed, his biographer Giorgio de Santillana stated that "It has been known for a long time that a major part of the church intellectuals were on the side of Galileo, while the clearest opposition to him came from secular ideas" (*The Crime of Galileo*, University of Chicago Press, 1955, xii-xiii).

2) Galileo (though correct in a general way) was overconfident and quite obstinate in proclaiming his scientific theory as absolute truth, and this was a major concern. Accordingly, St. Robert Bellarmine, who was directly involved in the controversy, made it clear that heliocentrism was not irreversibly condemned, and also that a not-yet proven theory was not an unassailable *fact*. In other words, he thought scientific theories and hypothesis were not dogmas, but provisional, and never absolutely proven (hence, Newton could be overthrown by Einstein and Planck and Heisenberg, etc.). Bellarmine didn't consider heliocentrism proven beyond all doubt, like Galileo did, and in that respect he was right. Bellarmine actually had the superior understanding of the nature of a scientific hypothesis, as Frederick Copleston, the famous Catholic historian of philosophy, has observed. Likewise, philosopher of science Thomas Kuhn, in his book, *The Copernican Revolution* (New York: Random House / Vintage Books, 1957, p. 226), after commenting on some folks who refused to look through Galileo's telescope, wrote:

> Most of Galileo's opponents behaved more rationally. Like Bellarmine, they agreed that the phenomena were in the sky but denied that they proved Galileo's contentions. In this, of course, they were quite right. Though the telescope argued much, it proved nothing.

Charles Darwin's "bulldog" Thomas Henry Huxley (see pages 177-181), no champion of the Catholic Church, also felt that the Church had actually made the better case, according to the level of scientific knowledge on the question of cosmology at that time.

3) True heliocentrism wasn't conclusively proven until some 200 years later, so Galileo's excessive confidence and near dogmatism about his own theory was misplaced. Galileo was overconfident in proclaiming his theory as fact, which is ironic, since the Church gets excoriated for doing the same thing from its perspective. Both sides were overly dogmatic. In essence, it was a case where one non-magisterial tribunal of the Church was

wrong about astronomy for (partially) the right reasons, and Galileo was partially right about astronomy for (partially) the wrong reasons.

4) Far more embarrassing and numerous "Bible vs. Science" fiascoes in the Protestant world are not nearly as well known. Martin Luther called Copernicus an "upstart astrologer" in 1539, appealing to Joshua 10:13 as proof that the sun moves. His successor Philip Melanchthon thought that Copernicus exhibited a lack of "honesty and decency," yet was an avid enthusiast of astrology. John Calvin "proved" geocentrism from Psalm 93:1, and contended that belief in a rotating earth would "pervert the order of nature." Francois Turretin, John Owen, and many Puritans followed suit. Catholic philosophers, on the other hand, like Nicholas Oresme (c.1325-1382) and Nicholas of Cusa (1401-1464) had long since posited a moving earth, and the sphericity of the earth had been taught even earlier by St. Albert the Great, St. Thomas Aquinas, and Dante. The Protestant University of Tubingen condemned the heliocentrism of the great Lutheran astronomer Johann Kepler (1571-1630), not long before the Galileo incident. Leibniz, the Lutheran philosopher (1646-1716) attacked Newton's theory of gravitation.

5) In 1633 Galileo was "incarcerated" in the palace of Niccolini, the ambassador to the Vatican from Tuscany, who admired Galileo, spent five months with Archbishop Piccolomini in Siena, and then lived in comfortable environments with friends for the rest of his life (though technically under "house arrest"). No evidence exists to prove that he was ever actually subjected to torture or deliberately blinded (he lost his sight in 1637).

6) One must counter the notion that the Galileo affair somehow proved that the Catholic Church is or was hostile to scientific inquiry. It is *not*, and this is quite clear, the more one studies the matter. Science, for its part, certainly has plenty of *its own* skeletons in the closet that we rarely hear about: such as widespread support for the racist pseudo-science of phrenology in the 19th and early 20th centuries: where the shape of a person's skull was thought (by mainstream science) to have a direct relationship to their intelligence. The science of, say, 1900, was shot through with racism: hardly a proud chapter in scientific

history. But Christians of two, three generations earlier, like William Wilberforce and the abolitionists were far more "progressive" on the race issue. Christians are always on the cutting edge of societal progress, whether we look at slavery, or civil rights (Dr. Martin Luther King: a Baptist pastor), or the fall of Soviet Communism (Pope John Paul II and Christians in Eastern Europe, and President Ronald Reagan). Eugenics is another sad chapter in scientific history. We saw what the Nazis did with that, with their experimentation on concentration camp inmates (Germany was a highly scientifically advanced nation). In America, there was sterilization of black men and suchlike. But this was supposedly "good science". Margaret Sanger picked it up and institutionalized her racism in her group, Planned Parenthood. The best Christian apologists of the period, like G. K. Chesterton and C. S. Lewis, wrote about these kinds of follies that were rampant within science. Lewis often satirized the tunnel vision materialist scientist of his time. Chesterton went after eugenics; both of them lambasted contraception, etc. There were also highly embarrassing escpades of folly such as so-called *Nebraska Man*: presented as compelling evidence for human evolution at the Scopes Trial in 1925: a single tooth that turned out to be that of an extinct pig; *Piltdown Man*, which was revealed as a (rather obvious) hoax in 1954 (after 42 years of being widely accepted and triumphalistically proclaimed as "evidence"), etc. It was standard belief for a long time that the universe was eternal, whereas we now know that is false, and Christians had always known it from revelation. And these were theories or "facts" touted by science proper. The Church (i.e., one of its tribunals) made a mistake (one not affecting the Catholic doctrine of infallibility); many prominent scientists and scientific institutions have made plenty of stupid mistakes too. But the point is not *merely* to note that scientists make mistakes (a thing anyone with a lick of sense knows), but rather, that Christians are not the *only* ones who make mistakes (specifically with the Galileo incident in mind) and that there are many aspects to the Galileo affair that many are unaware of. This is an exercise of pointing out double standards of presentation, by presenting (fairly) certain facts of history. Catholics got some things wrong in 17th century cosmology? So did everyone *else*, etc. Why, then (I would ask) are *Catholics* always discussed, and all this *other* scandalous history ignored and unknown? Catholic (and other Christian) mistakes are discussed forever and caricatured and distorted, but mistakes of either Galileo or science in general through the centuries are glossed-over, ignored, and it is pretended that there is this huge qualitative difference between Christian mistakes and any of the others.

7) Geocentrism was never taught nor infallibly defined by the Catholic magisterium (teaching authority). No one can produce any pope or official document where this was supposedly done.

8) The Danish scientist Tycho Brahe (1546-1601), whom Thomas Kuhn described in *The Copernican Revolution* as "the preeminent astronomical authority" of the second half of the 16th century, who had "immense prestige" and who was greater in "technical proficiency" than even Copernicus (p. 200), never became convinced of heliocentrism. Kuhn praised Tycho's "great ingenuity," "phenomenal achievements" of naked-eye observations, improvement of instruments and calculations, "the reliability and the scope of the entire body of data that he collected," and concluded, "Trustworthy, extensive, and up-to-date data are Brahe's primary contribution to the solution of the problem of the planets" (pp. 200-201). In other words, this man was an eminent scientist; the best astronomer, according to Kuhn, from 1550-1600 (only 16 or 33 years before Galileo's inquiries), despite the earlier Copernicus, and yet he, too, rejected heliocentrism. So then, why is it considered such a scandal that the Church (whose sphere of expertise is not science) was so outrageously wrong, for merely accepting a position similar to that of the scientist Tycho? One must have a proper historico-scientific perspective. Kuhn even claims about Tycho's cosmology (pp. 202-203):

> The Tychonic system is, in fact, precisely the equivalent mathematically to Copernicus' system. Distance determination, the apparent anomalies in the behavior of

the inferior planets, these and the other new harmonies that convinced Copernicus of the earth's motion are all preserved.

Also, Kuhn stated (p. 205):

. . . geometrically, the Tychonic and Copernican systems were identical.

Tycho, according to Kuhn, in fact introduced some innovations that went *beyond* Copernicus:

Brahe's system . . . forced his followers to abandon the crystalline spheres which, in the past, had carried the planets about their orbits . . . Copernicus himself had utilized spheres to account for the planetary motions . . . Any break with the tradition worked for the Copernicans, and the Tychonic system, for all its traditional elements, was an important break.
Brahe's skillful observations were even more important than his system in leading his contemporaries toward a new cosmology. They provided the essential basis for the work of Kepler, who converted Copernicus' innovation into the first really adequate solution of the problems of the planets . . . the new data collected by Brahe suggested the necessity of another major departure from classical cosmology -- they raised questions about the immutability of the heavens. (pp. 205-206)

9) Even the elliptical orbits and other crucial astronomical contributions of Johannes Kepler (1571-1630) were not immediately triumphant, as Kuhn notes:

[N]ot until the last decades of the seventeenth century did Kepler's Laws become the universally accepted basis for planetary computations even among the best practicing European astronomers.(p. 225)

10) A Catholic tribunal's condemnation of Galileo has nothing to do with Catholic infallibility. Therefore, the mistake made has no bearing on Catholic epistemological claims, since an error made in a non-infallible sense is simply that: an error: with no further implications for the Catholic system. How this determination is made is very simple:

A. It wasn't made by a pope.

B. It wasn't made by an ecumenical council (in agreement with, or ratified by a pope). Technically, neither papal and conciliar infallibility were expressly defined at that time at the highest levels of Catholic authority, though very widely believed and accepted by many centuries of practice; papal infallibility was made *ex cathedra* dogma in 1870 at the First Vatican Council; the Second Vatican Council treated conciliar infallibility in greater depth than ever before.

C. No accepted formula was expressed, in which all Catholic faithful were bound to hold this opinion as an article of the Catholic faith.

D. Even if *C* were true, the condemnation would be neither binding on all the faithful nor infallible, because of the source of the statement (per *A* and *B*).

E. It had nothing truly to do with faith and morals (Vatican I made clear that those areas alone were the subject of infallible declarations): meaning theology and ethical and moral issues. The astronomical details of the earth do not in any way constitute a point of theology.

This particular tribunal was obviously gravely mistaken in both its science and in its biblical hermeneutics, but not in theology, because this isn't a theological point in the first place. And since it is not (nor is it a moral matter), it cannot possibly be an *ex cathedra* or infallible statement. It's simply wrong, period, through and through, but this has no bearing on the Catholic doctrine of infallibility because the conditions for same were not met, per *A-E* above. In conclusion then, the attempt to enlist the Galileo affair as a "proof" of a profound internal consistency in Catholic ecclesiology and epistemology, fails, since it is fundamentally wrongheaded.

11) To generalize from one instance where mistakes were made, to "Christianity's view of science" or "irrational and anti-scientific religious thinking" is utterly absurd. One (non-infallible, non-magisterial) Catholic tribunal got it wrong! Why should it be such a big deal? It can be argued that this actually proves the fact that the Church is *not* opposed to science, since Galileo is the one "stock argument" trotted out *ad nauseum* (just as Popes Honorius, Vigilius, and Liberius are always trotted out to supposedly disprove papal infallibility). Atheists and agnostics (to turn the table) wouldn't concede for a second that Communism, Stalinism, Maoism, Naziism, eugenics, phrenology, astrology, alchemy, sterilization of black men, etc., were all indicative of "problems with atheist thinking" so that they would have to waste time defending atheism and atheists against these charges, as if such a broad generalization can be made in the first place. The overall historical picture has to be taken into account.

12) Galileo's errors are arguably more foolish than the Church's errors, insofar as he was dogmatic from the standpoint of the epistemology of science, where that has no place whatever. One *expects* religious bodies to be dogmatic by their very *nature*, because they claim to be conveying revealed truths of revelation. But dogma supposedly has no place in science (Thomas Kuhn and Stephen Jay Gould thought quite otherwise, insofar as how science is actually *practiced*). The Galileo tribunal wrongly interpreted the Bible as if it precluded either heliocentrism or a rotating earth. The Bible's not a science book and it has to be interpreted according to the principles of phenomenological description and anthropomorphism and anthropopathism. We do this ourselves, naturally, all the time, by saying "the sun rose at 5 AM" or "the stars moved across the sky." We can give Galileo's factual scientific errors a pass because he was early in the modern scientific scene. Science builds on the shoulders of past giants,

and at that time there weren't many "giants" in terms of *modern* scientific method. But on the very same basis, the Church of that time ought to *also* be given a "pass". Logically, those who wish to bash the Church for the error made with Galileo should direct at least equal (if not more) ire at Galileo himself for *his* errors. But it is inconsistent to blast the Church on the basis of errors made, while giving Galileo a complete pass.

## Chapter Twelve

## Galileo and Other Prominent 16th-17th Century Astronomers' Acceptance of Astrology

One of the most cherished myths of our secular culture is "Religion (Christianity) vs. Science" or "Reason vs. Faith" or "Science vs. the Bible" -- as if the two things are inexorably opposed, by their very nature. They are not at all, of course. And this is obviously the case, since science deals with physical matter and the causative laws affecting it, whereas religion deals primarily with non-material things such as ethics, soul, spirit, God the spirit, love, faith, and so forth.

In other words, basically, it is an "apples and oranges" scenario. The two (both rightly understood) need not clash at all; there is no necessity whatsoever for there to be any earth-shaking conflict (and there usually isn't). Galileo got it right when he said that "the Bible teaches us how to go to heaven; not how the heavens go." That is as good of a brief summary as anyone could give on the subject. Science is science; theology is theology. Is this not obvious? There is overlap, of course (as there is between religion and philosophy), but they are two essentially different fields of study or belief, in aims, subject matter, and methods.

Since the Renaissance, however, and the advent of (Baconian, Copernican, Newtonian) modern science and scientific method as a (rightly) cherished vehicle for arriving at truths concerning the laws of nature and this world and the universe, for some reason, a certain type of secular, skeptically inclined, usually religiously nominal or liberal mindset, including many (but not *all*, by *any* means) scientists want to keep the old "religion vs. science" false dichotomy going, for whatever reasons. Sometimes it is largely a reaction against the ongoing "creation vs. evolution" controversy.

Other times, it is an obsessive dwelling on one-time historical errors such as the Galileo fiasco or some of the nonsense that went down during the initial reaction to Darwin's *Origin of Species*, or the notorious Scopes Trial of 1925. Certain tiny, fringe factions on the Christian side have also adopted an "anti-science" or "know-nothing" attitude as well, which does no good.

Impressions formed by the incessant trumpeting of these unfortunate events apparently have a staying power for those otherwise inclined to question Christianity and the biblical revelation. Stereotypes thus arrived at have a dismaying influence over minds and serve a purpose of minimizing strains of thought (intellectually respectable Christian theology and faith) that are deemed by certain folks as intellectually retrograde, undesirable, indefensible, and excessively or completely irrational.

Part of the larger myth here considered is the notion of a sort of black-and-white dichotomy, as if the scientists in any given conflict with religion or faith (e.g., Darwin and Galileo), were these rationalistic, reasonable, always objective, solely truth-seeking machines, so to speak, whereas those on the Christian "side" were invariably dogmatic, closed to reason and inquiry and scientific observation, and indeed, *opposed* to same (hence we hear about the few Catholic throwbacks who refused to peer into Galileo's telescope or young-earth creationists with their various kooky "scientific" arguments for a 6000-year old earth).

There is also a sub-bias that operates within the larger secular vs. religion mentality; somewhat related to it: that of early Protestantism being more open to science than Catholicism was (based largely or disproportionately, I imagine, on the grossly exaggerated implications of the Galileo episode). For some people (including many Protestants, as one would expect), Protestantism is considered a bit more rational and closer to "intelligent, secular, scientific thought" than Catholicism is. They simply assume that it had -- historically -- a more open attitude towards science. This is *certainly* untrue concerning the period of the 16th and 17th centuries (if it ever was true at any time, which is highly debatable). It is only relatively small factions of both faiths that fell into the error of thinking in "anti-science" terms.

It is high time to turn our attention to the flaws and faults (in retrospect) of many of the 16th and 17th century scientists who are "used" for the purpose of bashing Christianity. I will be examining their connection to, and adherence of astrology: a thing that is now considered by virtually all scientists as false and unscientific (pseudo-science at best and sheer fabrication at worst), and scorned and rejected by Protestants and Catholics and Orthodox, too, as occultic nonsense (and often, in practice, quackery). In fact, astrology and astronomy, by all accounts, were very closely connected in the 16th century. It was difficult to disentangle one from the other.

The Christian / faith side isn't always completely "bad" or wrong in conflicts and the scientists aren't always free from error and dogmatism: even on a large scale. The acceptance of astrology illustrates this. If the Christians must be pilloried and mocked and ridiculed for their errors, then isn't it fair *also* to point out something like *this*: the widespread acceptance of astrology, even in scientific circles: among the very greatest scientists? What's good for the goose is good for the gander.

Yet what often happens is that we Christians usually have to "pay" for our past mistakes forever, while similar whoppers on the scientific side are soon forgotten and hardly ever brought up at all. The actual facts are far more fascinating. A great Catholic scientist like Galileo, in the forefront of the emerging heliocentrism, had a strong belief in astrology and wrote many astrological charts! Ggreat Lutheran scientists such as Kepler and Tycho Brahe were doing the same thing. Even Isaac Newton was fascinated with alchemy, a sort of half-sister to astrology. The categories of the myths don't fit; they don't work: the whole scenario gets turned upside down.

The best early scientists are not *supposed* to be good Christians, according to the stupid, anti-Christian myth, but almost all of them (like the later Gregor Mendel: the monk who founded genetics) *were*. Likewise, the scientists (who are the "good guys" and enlightened folks, over against the oppressive, know-nothing Catholic Church and the Lutherans) are not *supposed* to believe in something so absurd as astrology (or alchemy). But they *did*. So the myth falls and fails at least four times:

> 1. Science is not always absolutely right and religion absolutely wrong when they come into conflict. This is a gross simplifying of the historical reality.
>
> 2. Protestantism has not been -- historically; especially in the 16th and 17th centuries -- more open to, and less hostile to science than Catholicism. In fact, a good case might be made for the *contrary* position.
>
> 3. The great early scientists were usually Christians of some sort (this clearly suggests that the often trumped-up conflict is not inherent).
>
> 4. The great early scientists made huge mistakes just as the Christians have sometimes done (witness: astrology).

Conclusion: there is plenty enough error and folly to go around; therefore Christians of all stripes are often subjected to a quite unfair and outrageous "bum rap". The more knowledge we have of the relevant historical particulars, the better we understand this.

Historian of science Thomas S. Kuhn wrote in his book, *The Copernican Revolution* (New York: Random House / Vintage Books, 1957, 93-94):

> [A]strology was inseparably linked to astronomy for 1800 years; together they constituted a single professional pursuit . . . those who gained fame in one branch were usually well known in the other as well . . . European astronomers like Brahe and Kepler . . . were supported financially and intellectually because they were thought to cast the best horoscopes.
>
> During most of the period with which the rest of this book is concerned, astrology exercised an immense influence upon the minds of the most educated and cultured men of Europe . . . during the late Middle Ages and the Renaissance, astrology was the guide of kings and of their people, and it is no accident that these are just the periods during which earth-centered astronomy made most rapid progress . . . astrology became a particularly important determinant of the astronomical imagination.
>
> . . . It cannot be coincidence that astrology's stranglehold upon the human mind finally relaxed during just the period in which the Copernican theory first gained acceptance. It may even be significant that Copernicus . . . belonged to the minority group of Renaissance astronomers who did not cast horoscopes.

Georg Joachim Rheticus (1514-1574): Copernicus' sole student, published in 1540 in Danzig, *Narratio prima*, a work that defended both heliocentrism and astrology.

Tycho Brahe (1546-1601) was deeply immersed in astrology. He analyzed a supernova of 1572 and wrote an astrological report on it, entitled The New Star. He concluded that it was related to the preceding New Moon, ruled by Mars, and thought it was a harbinger of huge upheavals in politics and religion and indeed, a new age. He lectured on astrology at the

University of Copenhagen, and gave readings to his patron, King Frederick II. In an Internet article, ("Tycho Brahe and Astrology"), Adam Mosley observed:

> Like the fifteenth-century astronomer Regiomontanus, Tycho Brahe appears to have accepted astrological prognostications on the principle that the heavenly bodies undoubtedly influenced (yet did not determine) terrestrial events, . . . Two early tracts, one entitled *Against Astrologers for Astrology*, and one on a new method of dividing the heavens into astrological houses, were never published and are now lost. . . . An astrological world-view was fundamental to Tycho's entire philosophy of nature. His interest in alchemy, particularly the medical alchemy associated with Paracelsus, was almost as long-standing as his study of astronomy, and Uraniborg was constructed as both observatory and laboratory. In an introductory oration to the course of lectures he gave in Copenhagen in 1574, Tycho defended astrology on the grounds of correspondences between the heavenly bodies, terrestrial substances (metals, stones etc.), and bodily organs.

In another online article, "Tycho Brahe: A King Amongst Astronomers," David Plant states:

> He was an imperious, hard-drinking aristocrat whose devotion to precise observation was motivated by his devotion to astrology. He was also an alchemist and lived for twenty years on a fantastic 'sorcerer's island' near Hamlet's castle of Elsinore, ending his days as Imperial Mathematicus at the court of the Holy Roman Emperor. . . . Tycho also found time to provide an annual astrological almanac for King Frederick and to write detailed reports on the horoscopes of the king's children. The royal horoscopes were presented as handsome bound volumes with up to 300 pages of natal delineations and directions.

In the Internet paper, "Galileo's Astrology," Nick Kollerstrom provides many facts about the great astronomer's abiding interest in and advocacy of astrology:

> Galileo, like Kepler, was a *mathematicus*, a term which had a threefold meaning as referring to mathematics, astrology and astronomy. . . . some twenty-five charts drawn up by him do remain, plus several instances of his chart analyses. . . . His revolutionary bestseller, *Sidereus Nuncius*, 'The Message of the Stars' appearing in March, 1610, opened with an eloquent account of the traditional qualities assigned to Jupiter:
>
>> So who does not know that clemency, kindness of heart, gentleness of manners, splendour of royal blood, nobleness in public functions, wide extent of influence and power over others, all of which have fixed their common abode and seat in your highness - who, I say, does not know that these qualities, according to the providence of God, from whom all good things do come, emanate from the most benign star of Jupiter?
>
> . . . The text follows with an account of Jupiter's position at the top of the chart of his young patron, Cosimo de Medici, the Duke of Tuscany:
>
>> Jupiter, Jupiter I say, at the instant of Your highness's birth had already passed the slow, dull vapours of the horizon and was occupying the Midheaven, from which point it was illuminating the eastern angle, from that sublime throne saw the most happy delivery and all the splendour and magnificence of the newly-born diffused in the most pure air . . .
>
> . . . Galileo's text continued:

> . . . in order that your tender body and your mind might imbibe with their first breath that universal influence and power, . . .

. . . Galileo not only drew up charts for his two illegitimate daughters, but composed character-judgements based upon them. For Virginia the elder daughter he noted that the Moon (traditionally of feminine significance for motherhood, etc) was 'debilitated', and wrote grimly:

> The Moon is very debilitated and in a sign which obeys. She is dominated by family relationships. Saturn signifies submission and severe customs which gives her a sad demeanour, but Jupiter is very well with Mercury, and well-aspected corrects this. She is patient and happy to work very hard. She likes to be alone, does not talk too much, eats little with a strong will but she is not always in condition and may not fulfil her promise.

In the younger daughter Livia he discerned . . . a more extrovert character. Her *De Ingenio* affirmed:

> Mercury rising is very strong for all things, and Jupiter which is conjunct gives knowledge and bounty, simplicity, humanity, erudition and prudence.

. . . French philosophers such as Descartes and Gassendi were sceptical towards astrology, whereas this had not become an issue in Renaissance Italy: there was no social context as could have supported astronomers sceptical towards astrology during Galileo's life. Only later on, in the latter half of the seventeenth century, was astrology expelled from the universities, whereby astronomy became established as a separate and independent discipline.

Johannes Kepler (1571-1630), the great astronomer, was also a fervent devotee of astrology. Sachiko Kusukawa provides some basic facts:

> As one historian, John North, put it, 'had he not been an astrologer he would very probably have failed to produced his planetary astronomy in the form we have it.'
> Kepler believed in astrology in the sense that he was convinced that planetary configurations physically and really affected humans as well as the weather on earth. He strove to unravel how and why that was the case and tried to put astrology on a surer footing, which resulted in the *On the more certain foundations of astrology* (1601). In *The Intervening Third Man, or a warning to theologians, physicians and philosophers* (1610), posing as a third man between the two extreme positions for and against astrology, Kepler advocated that a definite relationship between heavenly phenomena and earthly events could be established.
> At least 800 horoscopes drawn up by Kepler are still extant, several of himself and his family, accompanied by some unflattering remarks. As part of his duties as district mathematician to Graz, Kepler issued a prognostication for 1595 in which he forecast a peasant uprising, Turkish invasion and bitter cold, all of which happened and brought him renown. Kepler is known to have compiled prognostications for 1595 to 1606, and from 1617 to 1624. As court mathematician, he explained to Rudolf II the horoscopes of the Emperor Augustus and Mohammed, and gave astrological prognosis for the outcome of a war between the Republic of Venice and Paul V. In the *On the new star* (1606) Kepler explicated the meaning of the new star of 1604 as the conversion of America, downfall of Islam and return of Christ. The *De*

*cometis libelli tres* (1619) is also replete with astrological predictions.

In his article, "Kepler and the Music of the Spheres," David Plant gives further detail:

> . . . we can at least begin by dismissing the notion that he rejected astrology out-of-hand. In the official history of scientific progress, the values of the Age of Reason and Industrial Revolution were projected onto the brilliant mathematician who had unravelled the laws of planetary motion. It seemed inconceivable that he could be tainted with the medieval superstition of astrology. Like Isaac Newton's passion for alchemy and theology, this aberration was best glossed over or, as actually happened in Kepler's case, twisted into a distortion of the truth.
> . . . He was always careful to distinguish his reverential vision of the celestial harmonies from the practices of the backstreet astrologers and almanac-makers "who prefer to engage in mad ravings with the uneducated masses". Kepler's astrology was on another plane altogether. . . . Kepler was neither the first nor the last astrologer to pour scorn on those who practise apparently inferior forms of the art. His disapproval stems from his conviction that astrology is nothing less than a divine revelation, ". . . a testimony of God's works and... by no means a frivolous thing".
> . . . From his long-term study of weather conditions correlated with planetary angles and from detailed analysis of his collection of 800 birth charts, Kepler concluded that when planets formed angles equivalent to particular harmonic ratios a resonance was set up, both in the archetypal 'Earth-soul' and in the souls of individuals born under those configurations. He considered this 'celestial imprint' more important than the traditional emphasis on signs and houses: "in the vital power of the human being that is ignited at birth there glows that remembered image . . ."

Even Sir Isaac Newton (1642-1727) was heavily involved in a somewhat related field (alchemy) that is now completely rejected and scorned by science. It is known that he possessed 169 books on the topic (or 9.6% of his personal library; see: J. Harrison, *The Library of Isaac Newton*, Cambridge University Press, Cambridge, 1978, 58-78). Sue Toohey writes, in her research article, "Isaac Newton and the Ocean of Truth":

> Throughout his life, Newton spent more time intensely involved with alchemy than any of his scientific pursuits. Many of his biographers, confronted with what they see as completely divergent writings from Newton, have chosen to gloss over anything that does not fit easily into the image of Newton generally acknowledged. Anything that has not been considered in keeping with his scientific discoveries has often been regarded as misguided.
>
> . . . Newton, like most alchemists of the time, believed that alchemic wisdom extended back to ancient times. He believed strongly in the religious and astrological symbolism of alchemy. Most alchemists of the day were adept at astrology, sharing much of the deeper symbolism of the two disciplines, including the connection between the seven metals and the seven planets, as well as the four elements and the four humours. Newton became involved in secretive alchemical networks, devoting time to copying out the unpublished alchemical treatise passed around among them. . . . Newton often pleaded with fellow alchemist Robert Boyle to keep silent in publicly discussing alchemy. But, rather than being uncomfortable with his participation in alchemy, it seems that Newton believed that this secret knowledge was not for everyone. He felt that the Hermetic writers of the past had concealed their work for good reason and Newton was prepared to honour this adherence to secrecy.
>
> . . . Until the late seventeenth century, almost all astronomers were astrologers. Spencer sees that modern

astronomy's contempt for its mystically minded ancestor has required an acrobatic rewrite of history, in which the ideas of those of the past have been bowdlerised and suppressed. Nowhere is this more evident than in the case of Isaac Newton. When he died on 20 May 1727, those seeking to portray Newton as a rationalist rejected most of his non-scientific works. They remained unknown for over two hundred years.

It should be noted in conclusion that the Lutheran astronomer Michael Maestlin (1550-1631), opposed astrology, as did Catholic astronomers Pierre Gassendi (1592–1655) and Giovanni Domenico Cassini (1625–1712).

## Chapter Thirteen

## "No One's Perfect": Scientific Errors of Galileo and 16th-17th Century Cosmologies Rescued from Inexplicable Obscurity

For some reason many of the more loudmouthed and absurdly overconfident advocates of (what they consider essentially materialistic) science and/or critics of Christianity (who endlessly bring up *and* distort the Galileo affair) are reluctant to admit that there is more than enough historic scientific error (hindsight is 20/20) to go around.

Most Catholics in that early period of modern astronomy didn't get everything right, but neither did *anyone else* (including even the best scientists) get even some very basic facts of astronomy right. So why is one party excoriated, while the errors of the vaunted (and indeed brilliant) scientists are ignored, unknown, or suppressed, in a cynical effort at one-sided presentation?

The objective observer will note, I submit (upon a complete perusal of the relevant facts), that in most cases of supposed stark opposition of two competing ideas (especially ones as complex as those involved in science and philosophy), there is truth and error to be found on both sides. The reality of various conflicts in the realm of the history of ideas is not usually "good vs. evil." Just as individuals are radical mixtures, so are sets of ideas: with some falsehood mixed in. Let me present, if I may, some basic facts:

**Nicolaus Copernicus** (1473-1543) erred in asserting circular orbits and in holding that the sun was the stationary center of the *universe*, with not only the earth and the other planets of the solar system, but also *all* the other stars, moving around it. He also believed that transparent rotating crystalline spheres carried the planets in their orbits.

**Tycho Brahe** (1546–1601) erred insofar as he was a geocentrist and held (Tychonic "geoheliocentric" system) that the sun and moon revolve around the earth, and the other five planets revolve around the sun: all in circular -- not elliptical -- orbits. Also, in his system the earth did not rotate.

**Johannes Kepler** (1571-1630) was correct in asserting elliptical orbits of the planets around the sun, at varying speeds (both notions having been foreseen by the Catholic Cardinal Nicholas of Cusa in the 15th century), but continued to err in thinking that the sun was the center of the entire universe. The idea that the sun was but one of innumerable stars, was strongly advocated by the mystic heretic and scientist Giordano Bruno (1548-1600). According to the Wikipedia entry, Bruno understood several aspects of cosmology that even Copernicus, Kepler, Galileo, and Tycho neglected to see:

> Bruno believed ... that the Earth revolves around the sun, and that the apparent diurnal rotation of the heavens is an illusion caused by the rotation of the Earth around its axis. Bruno also held (following Nicholas of Cusa) that because God is infinite the universe would reflect this fact in boundless immensity. Bruno also asserted that the stars in the sky were really other suns like our own, around which orbited other planets. ...
>
> Bruno's infinite universe was filled with a substance—a "pure air," aether, or *spiritus* -- that offered no resistance to the heavenly bodies which, in Bruno's view, rather than being fixed, moved under their own impetus. Most dramatically, he completely abandoned the idea of a hierarchical universe. The Earth was just one more heavenly body, as was the Sun. ...
>
> Under this model, the Sun was simply one more star, and the stars all suns, each with its own planets. Bruno saw a solar system of a sun/star with planets as the fundamental unit of the universe. According to Bruno, infinite God necessarily created an infinite universe, formed of an infinite number of solar systems, separated by vast regions full of Aether, because empty space could not exist. (Bruno did not arrive at the concept of a galaxy.)

**Galileo** (1564-1642) disbelieved in Kepler's elliptical orbits of the planets, considering the circle the "perfect" shape for planetary orbits:

> Galileo's two main published works were *Dialogue Concerning the Two Chief World Systems* in 1629 and *Discourses and Demonstrations Concerning Two New Sciences* in 1638. The first of these was fully ten years after Kepler published his third law of planetary motion, and twenty years after the publication of Kepler's first and second laws, yet Galileo seemed oblivious to those developments – despite the fact that he was very familiar with Kepler's works and had high regard for him

(referring to him as "a person of independent genius"). Einstein described Galileo's failure to take account of Kepler's laws as "a grotesque illustration of the fact that creative individuals are often not receptive".

(source: "Math Pages")

R. R. Reno referred to this error on 26 July 2010, on the blog *First Thoughts* (connected with the magazine *First Things*):

> These days no educated person "acknowledges" Galileo's heliocentric theory as "correct." Galileo adopted Copernicus's theory, which presumed lovely circular orbits, but that turns out to be wrong. Tycho Brahe painstakingly collected data about the positions of the planets in the sky, which was theorized by Johannes Kepler as eliptical rather than circular motion.
> Interestingly, Kepler and Galileo corresponded, but Galileo insisted on defending Copernicus' views. On this point, Galileo was mistaken, and not just because he did not have access to the scientific data and good arguments. He was, like many brilliant individuals, a vain and willful man.

Scott Rosmarin, in his article, "Galileo's Lapse - The Fallibility of Scientists" (29 March 2010), noted:

> Johannes Kepler had provided plausible evidence that the planets move in *elliptical,* nor circular orbits, and *not* at uniform speeds, but *variable* speeds, depending on their distance from the sun. This seriously challenged the Copernican view. Galileo . . . simply rejected Kepler's view, clinging instead to the ancient belief that circular motion was "beautiful" and, therefore, privileged. . . . Galileo believed dogmatically in the Copernican view, not merely as a good starting hypothesis, or true subject to possible modifications, such as those offered by Kepler.

Galileo was also wrong in following Copernicus's (and Kepler's) view that the sun was the stationary center of the universe, with the earth and other planets of the solar system, and also *all the other stars*, moving around it. In this respect, he and Copernicus had hardly advanced beyond what was already posited by the ancient Greek astronomer Aristarchus (d. c. 230 B. C.). All three had merely moved the center of the universe 93 million miles from the earth, to the sun.

That is not all that different (knowing how large the universe is) from positing that the earth is the center. Both are vastly erroneous positions. But, oddly enough, we only hear about one error and not the other. Nicholas of Cusa (a Catholic Cardinal) and Giordano Bruno were closer to the truth in these respects than Copernicus, Galileo, and Kepler. Truth is stranger than fiction.

Galileo, moreover, argued vehemently in his 1623 book *The Assayer* that the comets of 1618 were merely an optical illusion. The Wikipedia entry on the book states:

> The book was a polemic against the treatise on the comets of 1618 by Orazio Grassi, a Jesuit mathematician at the Collegio Romano. In this matter Grassi, for all his Aristotelianism, was right and Galileo was wrong. Galileo incorrectly treated the comets as a play of light rather than as real objects. . . .
>
> Although *The Assayer* contains a magnificent polemic for mathematical physics, ironically its main point was to ridicule a mathematical astronomer. This time, the target of Galileo's wit and sarcasm was the cometary theory of a Jesuit, Orazio Grassi, who argued from parallax that comets move above the Moon. Galileo mistakenly countered that comets are an optical illusion.

The Wikipedia article, "Comet," observed that Galileo "rejected Tycho's parallax measurements and held to the Aristotelian notion of comets moving on straight lines through the upper atmosphere."

Furthermore, Galileo dismissed as a "useless fiction" the idea, held by his contemporary Johannes Kepler, that the moon caused the tides. He thought they were caused by the rotation of the earth. The *Stanford Encyclopedia of Philosophy* entry on Galileo comments on this notion and how it figured in the overall picture:

> This argument, about the tides, Galileo believed provided proof of the truth of the Copernican theory. . . . Galileo argues that the motion of the earth (diurnal and axial) is the only conceivable (or maybe plausible) physical cause for the reciprocal regular motion of the tides. He restricts the possible class of causes to mechanical motions, and so rules out Kepler's attribution of the moon as a cause. How could the moon without any connection to the seas cause the tides to ebb and flow? Such an explanation would be the invocation of magic or occult powers. So the motion of the earth causes the waters in the basins of the seas to slosh back and forth, and since the earth's diurnal and axial rotation is regular, so are the periods of the tides; the backward movement is due to the residual impetus built up in the water during its slosh. Differences in tidal flows are due to the differences in the physical conformations of the basins in which they flow . . . .
> 
> One can see why Galileo thinks he has some sort of proof for the motion of the earth, and therefore for Copernicanism. Yet one can also see why Bellarmine and the instrumentalists would not be impressed. First, they do not accept Galileo's restriction of possible causes to mechanically intelligible causes. Second, the tidal argument does not directly deal with the annual motion of the earth about the sun. And third, the argument does not touch anything about the central position of the sun or about the periods of the planets as calculated by Copernicus.

## Chapter Fourteen

## The Execution of Antoine Lavoisier: the Great Catholic Scientist and "Father of Chemistry" by "Enlightened" French Revolutionaries

We always hear, *ad nauseum*, about Galileo and how he was persecuted by the Catholic Church for saying that the earth went around the sun. It makes for great copy, because it was a whopper of a scientific error, for a tribunal of the Church to defend geocentrism, and because it is fodder for anti-Catholicism and contra-Catholicism, and even a broader anti-Christianity (and sometimes anti-theism and anti-religion, period). When Catholics make mistakes, they are never forgotten, and milked (in distorted, half-baked form) for all they are worth, for literally hundreds of years.

Atheists, agnostics, and secularists, who have a strong tendency to replace Christianity with scientism, as their new "religion" and worldview, absolutely love the Galileo story. It's been wonderful propaganda for them for almost 500 years; that is, in its usual modified, revisionist, one-sided presentation, designed to make the Catholic Church look as bad as it can possibly look, in order to foster the myth that it is somehow "anti-science."

But this is not only untrue; it is about as untrue to the facts of history as anything can *imaginably* be (as I am presently trying to repeatedly demonstrate).

Now, with that background in mind, let's consider *another* (far less well-known) case of a very renowned, skilled, important scientist being "persecuted" and wrongly treated. To give some biographical background, Antoine Lavoisier (1743-1794) is considered the "father of modern chemistry". He stated the first version of the law of conservation of mass, helped construct the metric system, wrote the first extensive list of elements, and helped to reform chemical nomenclature. He discovered that, although matter may change its form or shape, its mass always remains the same.

He determined that the components of water were oxygen and hydrogen, and that air was a mixture of gases, primarily nitrogen and oxygen. His *Traité Élémentaire de Chimie* (*Elementary Treatise on Chemistry*, 1789) is considered to be the first modern chemistry textbook. This text clarified the concept of an element as a substance that could not be broken down by any known method of chemical analysis, and presented a theory of the formation of chemical compounds from elements. Lavoisier introduced the possibility of allotropy in chemical elements when he discovered that diamond is a crystalline form of carbon. [source: Wikipedia biography]

Unfortunately, Lavoisier died prematurely at age 50. But how many know *how* he died (I just recently learned this, myself). He was beheaded by a guillotine: killed by the folks in France who prided themselves (and are widely known, for some bizarre reason, to this day), as proponents of "enlightened" reason: freed (so they viewed themselves) from the shackles of centuries of Christian "Dark Ages" intellectual slavery and mindless dogmatism (as the stereotype goes).

These were allegedly the "smart" people; the "liberated and free" ones; the "liberals" with all the noble ideas, fighting entrenched, decrepit tradition. "Liberty, equality, and fraternity," and all that. Even Thomas Jefferson admired them -- but at length he had to reluctantly concede that his friend John Adams had the superior (opposed) view as to the French Revolution.

But it is extremely interesting that we rarely hear of this epoisode in the history of science, and the response of various societies and worldviews to it. Most people, I suspect, are like myself. I had at least heard of the man's name, and knew he had a significant place in the history of scientific pioneers and great minds, but I didn't know he died in *this* fashion. And that is because the powers that be: the secularists who run large portions of our educational system, and those who dominate academia and our colleges, have a vested interest in downplaying and obscuring inconvenient facts such as these.

Their goal is to pit religion *against* science, so Galileo fits right in with that. They're *not* so keen on letting it be made known that Lavoisier, the father of chemistry, was murdered by a band of "enlightened" French revolutionaries: intent on tearing down tradition and especially the Church, and replacing Christianity with the "goddess of reason." *That* doesn't fit with the plan! So few know about it.

It's supposed to always be "Christianity against science" and never "radical secularism and atheism against science." That won't do. It doesn't fit in with the prevalent mythology and legend-building. The Church vs. Galileo the great Catholic scientist is great copy and trumpeted from the rooftops. But the atheist French revolutionaries against Lavoisier the great Catholic scientist is a tidbit of history that is ignored, suppressed and hushed up. It's a classic case of cynical, deliberate historical and academic bias.

What do we know about what happened to Lavoisier? What is *his* story (that deserves to be heard and learned about)? First of all, he died a Catholic. Grimaux, the first biographer to have access to his personal papers, wrote:

Raised in a pious family which had given many priests to the Church, he had held to his beliefs. To Edward King, an English author who had sent him a controversial work, he wrote, 'You have done a noble thing in upholding revelation and the authenticity of the Holy Scripture, and it is remarkable that you are using for the defence precisely the same weapons which were once used for the attack.'

(*Catholic Encyclopedia*: "Antoine-Laurent Lavoisier")

Lucien Scheler and W. A. Smeaton wrote a paper entitled, "An account of Lavoisier's reconciliation with the church a short time before his death," published in *Annals of Science*, Volume 14, Issue 2, June 1958, pp. 148-153. He received Holy Communion [source].

How was he treated at *his* trial? What was the basis of his "guilt"? Galileo spent his last several years under a very mild "house arrest" in comfortable environs, including palaces, and wasn't forbidden from carrying on his scientific experiments. Lavoisier was not quite so graciously treated:

> Early in his career he felt the need of increasing his resources to meet the necessities caused by his scientific experiments. With this in view he became a deputy *fermier-général*, whereby his income was much increased. But joining this association of State-protected tax-collectors only prepared the way for many years of bitter attack and a share of the public odium attaching to their privilege. He headed many public commissions requiring scientific investigation, he aimed at bringing France to such a state of agricultural and industrial expansion that the peasant and the working-man would have profitable employment and the small landed proprietor relief from the burdensome taxes hitherto purposely increased to make grants to corrupt favourites of the Court. Having incurred the hatred of Marat he

found himself, together with his fellow *fermiers-général*, growing more and more unpopular during the terrible days of the Revolution. Finally in 1794 he was imprisoned with twenty-seven others. A farcical trial speedily followed, in which he was charged with "incivism" in that he had damaged public health by adding water to tobacco. He and his companions, amongst them Jacques Alexis Paulze, his father-in-law, were condemned to death.

(*Catholic Encyclopedia*: "Antoine-Laurent Lavoisier")

Jean-Paul Marat, who made himself Lavoisier's enemy: one of the three leading figures of the French Revolution (along with Georges Danton and Maximilien Robespierre), was a failed scientist, and his motivations were clearly petty jealousy, and a wounded pride, fed by paranoia. And so he wrote:

> This persecution began at the moment the Academy realized that my discoveries about the nature of light upset its own work. . . . Since the d'Alamberts, the Condorcets, the Moniers, Monges, Lavoisiers and all the other charlatans of that scientific body wanted to hog the limelight for themselves . . . it isn't difficult to understand why they disparaged my discoveries throughout Europe, turned every learned society against me and had all learned publications closed to me.

(cited by Joe Jackson, *A World on Fire: A Heretic, an Aristocrat, and the Race to Discover Oxygen* [Penguin, 2007], p. 268)

For much more on Marat and his charades and shenanigans, see Jackson, *ibid.*, pp. 267 ff. (accessible on Google Reader). But Marat was actually already dead by the time Lavoisier was executed: having been stabbed to death in his bathtub by Charlotte Corday: a woman from a rival faction. He wound up like Jim Morrison (in a bathtub in Paris): except that his heart gave out due to a knife, rather than drugs and alcohol.

Marat himself was not, however, an atheist. He was, according to biographer Ernest Belefort Bay, a proponent of "Rousseauite Deism" and "vague Deism" (*Jean-Paul Marat - The People's Friend* [Vogt Press, 2008], pp. 84-85). He actually *persecuted* atheists:

> As long as the atheist only reasons, let him live in peace; but when, instead of keeping himself to the sceptical attitude, he declaims, when he dogmatises, when he seeks to obtain proselytes, becoming from that moment sectarian, he makes a dangerous use of his liberty, and he ought to lose it. Let him then be shut up for a limited time in a humane gaol.
>
> (Bay, *ibid.*, p. 85)

As usual, one whom atheists claim as one of their own was not (like Hume, Einstein, and many others) actually an atheist. Historical revisionism abounds. But Marat received the royal treatment when he died:

> Jacques-Louis David took up the task of immortalizing Marat in the painting *The Death of Marat*, . . . His heart was embalmed separately and placed in an urn in an altar erected to his memory . . . On 19 November, the port city of Le Havre-de-Grâce changed its name to Le Havre-de-Marat and then Le Havre-Marat. When the Jacobins started their dechristianisation campaign to set up the *Cult of Reason* of Hébert and Chaumette and *Cult of the Supreme Being* of Robespierre, Marat was made a quasi-saint, and his bust often replaced crucifixes in the former churches of Paris.
>
> (Wikipedia, "Jean-Paul Marat")

Joe Jackson described Lavoisier's witnessing of the pathetic hysterics and hero-worship at Marat's funeral:

> Lavoisier . . . must have sighed in relief, thinking that his foe could not persecute him any longer. But then, as a member of the National Guard, he was obliged to stand at attention during Marat's funeral . . . the painter David, the ceremony's organizer, clad Marat in a toga and crowned him with laurels. His body rested on a raised couch, drawn by twelve men; young girls surrounded the couch, all dressed in white and carrying wands and branches of cypress. The grief for the "martyr Marat" was as violent as the jeers for his murderess. "Oh heart of Jesus!" people sobbed along the route, some falling to their knees. "Oh sacred heart of Marat!"

(Jackson, *ibid.*, p. 287)

Lavoisier was one of the estimated 16-17,000 people (Jackson, p., 288) who were guillotined in the "enlightened" Reign of Terror, from October 1793 to July 1794. Here we have an atheist / deist / secularist / anti-Catholic Inquisition. We all know how incredibly tolerant the French Revolutionaries were. They harbored a special and tender love for the Catholic Church:

> Another anti-clerical uprising was made possible by the installment of the Revolutionary Calendar on 24 October. Hébert's and Chaumette's atheist movement initiated a religious campaign in order to dechristianize society. The program of dechristianization waged against Catholicism, and eventually against all forms of Christianity, included the deportation or execution of clergy; the closing of churches; the rise of cults and the institution of a civic religion; the large scale destruction of religious monuments; the outlawing of public and private worship and religious education; the forced abjuration of priests of their vows and forced marriages of the clergy; the word "saint" being removed from street names; and the War in

the Vendée. The enactment of a law on 21 October 1793 made all suspected priests and all persons who harbored them liable to death on sight. The climax was reached with the celebration of the goddess "Reason" in Notre Dame Cathedral on 10 November. Because dissent was now regarded as counterrevolutionary, extremist *enragés* such as Hébert and moderate Montagnard *indulgents* such as Danton were guillotined in the Spring of 1794. On 7 June Robespierre, who favoured deism over Hébert's atheism and had previously condemned the *Cult of Reason*, advocated a new state religion and recommended that the Convention acknowledge the existence of God. On the next day, the worship of the deistic *Supreme Being* was inaugurated as an official aspect of the Revolution. Compared with Hébert's somewhat popular festivals, this austere new religion of Virtue was received with signs of hostility by the Parisian public.

(Wikipedia, "Reign of Terror")

This was the lunacy that Lavoisier (a *real* man of reason) found himself tragically caught up in. It is estimated that some 70-72% of the victims were from the peasant working class, while 8% were nobles, 6% clergy, and 14% bourgeoisie (how ironic, but not surprising at all to those who know anything about how revolutions proceed).

There is no such thing as an anti-Catholic revolution without extreme hypocrisy involved (the butcher-tyrant Henry VIII's so-called "reformation" in England being the most obvious precedent). Here is how the great man of science was treated by the "enlightened" ones:

> One of twenty-eight French tax collectors and a powerful figure in the deeply unpopular Ferme Générale, Lavoisier was branded a traitor during the Reign of Terror by French Revolutionists in 1794. Lavoisier had also intervened on behalf of a number of foreign-born scientists including mathematician Joseph Louis

Lagrange, granting them exception to a mandate stripping all foreigners of possessions and freedom. . . .

An appeal to spare his life so that he could continue his experiments was cut short by the judge . . .

Lavoisier's importance to science was expressed by Lagrange who lamented the beheading by saying: *"Cela leur a pris seulement un instant pour lui couper la tête, mais la France pourrait ne pas en produire une autre pareille en un siècle."* ("It took them only an instant to cut off his head, but France may not produce another such head in a century.")

One and a half years following his death [on 8 May 1794], Lavoisier was exonerated by the French government. When his private belongings were delivered to his widow, a brief note was included reading "To the widow of Lavoisier, who was falsely convicted."

(Wikipedia, "Antoine Lavoisier")

The judge was Jean-Baptiste Coffinhal, who met his own death in the same fashion just three months later (on 6 August 1794). This seems to have been the end of so many of these fanatics, including Robespierre (28 July 1794) and Danton (5 April 1794). He who lives by the sword will die by the sword (someone said). Lavoisier had refused poison in prison, saying:

> I set no more value on life than you do; and why seek death before its time? It will have no shame for us. Our true judges are neither the tribunal that will condemn us nor the populace that will insult us. We are stricken down by the plague that is ravaging France. [source]

Biographer Arthur Donovan observed:

> Two centuries of additional investigation have still not turned up any evidence that Lavoisier was guilty of misconduct in the discharge of his many public duties.

(*Antoine Lavoisier: Science, Administration, and Revolution* [Cambridge Univ. Press, 1996], p. 296)

Philippe-Frédéric de Dietrich, fellow chemist, metallurgist, and associate member of the Academy of Science, was also killed in the Terror, on 19 November 1793. The famous French philosopher and mathematician Nicolas de Condorcet died in an "Enlightenment" prison under mysterious circumstances, on 28 March 1794. Jean Baptiste Gaspard Bochart de Saron, an astronomer and mathematician, fell prey to the terror on 20 April 1794. Guillaume-Chrétien de Lamoignon de Malesherbes, a botanist and statesman, met his end in the usual fashion, on 23 April 1794. Félix Vicq d'Azyr, a French physician and anatomist, originator of comparative anatomy and discoverer of the theory of homology in biology, died on 20 June 1794. His death may have had some relation to the Terror as well.

1793-1794 was a real banner period for French science and learning, and for Western Civilization. But all we ever hear about is Galileo . . .

# Chapter Fifteen

# Christian Influence on Science: Master List of Scores of Bibliographical and Internet Resources

## Bibliographical Sources (Books)

John Hedley Brooke, *Science and Religion: Some Historical Perspectives* (Cambridge, 1991).

John Hedley Brooke & Christopher Southgate, *God, Humanity and the Cosmos: A Textbook in Science and Religion* (T. & T. Clark Publishers, 2nd ed., 2005).

John Hedley Brooke and Geoffrey Cantor, *Reconstructing*

*Nature: The Engagement of Science and Religion* (T. & T. Clark, 1998).

Herbert Butterfield, *The Origins of Modern Science* (Free Press, rev. ed., 1997).

Thomas Cahill, *Mysteries of the Middle Ages: The Rise of Feminism, Science, and Art from the Cults of Catholic Europe* (Nan A. Talese, 2006).

Heidi Campbell and Heather Looy, editors, *A Science and Religion Primer* (Baker Academic, 2009).

Bernard Cohen, editor, *Puritanism and the Rise of Modern Science: The Merton Thesis* (Rutgers Univ. Press, 1990).

Marshall Clagett, *The Science of Mechanics in the Middle Ages* (Univ. of Wisconsin Press, 1959).

*Francis S. Collins*, *Language of God: A Scientist Presents Evidence for Belief* (Free Press, 2007).

A. C. Crombie, *Medieval and Early Modern Science* (two volumes, Doubleday Anchor, 1959)

A. C. Crombie, *Robert Grosseteste and the Origins of Experimental Science 1100-1700*, (Oxford: Clarendon Press, 1971).

A. C. Crombie, *The History of Science From Augustine to Galileo* (Dover Pub., 1996).

Richard C. Dales, *The Scientific Achievement of the Middle Ages* (Univ. of Pennsylvania Press, 1973).

Tihomir Dimitrov, editor, *50 Nobel Laureates and Other Great Scientists Who Believe in God* (online book, 2008)

Elaine Howard Ecklund, *Science vs. Religion: What Scientists Really Think* (Oxford Univ. Press, 2010).

Nancy K. Frankenberry, editor, *The Faith of Scientists: In Their Own Words* (Princeton Univ. Press, 2008).

Amos Funkenstein, *Theology and the Scientific Imagination from the Middle Ages to the Seventeenth Century* (Princeton Univ. Press, 1989).

Karl Giberson, *Worlds Apart: The Unholy War Between Religion and Science* (Beacon Hill Press, 1993).
Michael Allen Gillespie, *The Theological Origins of Modernity* (Univ. of Chicago Press, 2009).

Thomas F. Glick, Steven Livesey, and Faith Wallis, editors, *Medieval Science, Technology, and Medicine: An Encyclopedia (Routledge Encyclopedias of the Middle Ages)* (Routledge, 2005).

Edward Grant, *The Foundations of Modern Science in the Middle Ages: Their Religious, Institutional and Intellectual Contexts* (Cambridge, 1996).

Edward Grant, *God and Reason in the Middle Ages* (Cambridge, 2001).

Dan Graves, *Scientists of Faith: 48 Biographies of Historic Scientists and Their Christian Faith* (Kregel Resources, 1996).

Dan Graves, *Doctors Who Followed Christ: 32 Biographies of Historic Physicians and Their Christian Faith* (Kregel Pub., 1999).

James Hannam, *God's Philosophers: How the Medieval World Laid the Foundations of Modern Science* (Icon Books, 2010).

Peter Harrison, *The Bible, Protestantism, and the Rise of Natural*

*Science* (Cambridge Univ. Press, 2001).

R. Hooykaas, *Religion and the Rise of Modern Science* (Regent College Pub., 2000).

Toby Huff, *The Rise of Early Modern Science: Islam, China and the West* (Cambridge, 1993).

Charles E. Hummel, *The Galileo Connection: Resolving Conflicts Between Science and the Bible* (InterVarsity Press, 1986).

J. Wentzel Vrede van Huyssteen, editor, *Encyclopedia of Science and Religion* [online] (Macmillan: 2nd ed., 2003).

Stanley L. Jaki, *Cosmos and Creator* (Scottish Academic Press, 1981).

Stanley L. Jaki, *Science and Creation* (Scottish Academic Press, 1974).

Thomas Kuhn, *The Copernican Revolution* (New York: Vintage Books / Random House, 1959).

David Lindberg, editor, *Science in the Middle Ages* (Univ. of Chicago Press, 1978).

David Lindberg, *The Beginnings of Western Science* (Univ. of Chicago Press, 2nd ed., 2008).

David Lindberg and Robert Westman, editors, *Reappraisals of the Scientific Revolution* (Cambridge, 1990).

David Lindberg and Ronald Numbers, editors, *God and Nature: Historical Essays on the Encounter between Christianity and Science* (Univ. of California Press, 1986).

David Lindberg and Ronald Numbers, editors, *When Science and*

*Christianity Meet* (Univ. of Chicago Press, 2003).

Donald M. MacKay, *Science, Chance and Providence* (Oxford Univ. Press, 1978).

Donald M. MacKay, *Open Mind and Other Essays* (InterVaristy Press, 1988).

Alister E. McGrath, *Science and Religion: A New Introduction* (Wiley-Blackwell, 2nd ed., 2009).

J. P. Moreland, *Christianity and the Nature of Science: A Philosophical Investigation* (Baker Books, 2nd ed., 1999).

Robert P. Multhof, *The Origins of Chemistry* (F. Watts, 1967).

Ronald L. Numbers, editor, *Galileo Goes to Jail and Other Myths About Science and Religion* (Harvard Univ. Press, 2009).

Arthur Peacocke, *Creation and the World of Science: The Re-Shaping of Belief* (Oxford Univ. Press, 2nd ed., 2004).

John C. Polkinghorne, *The Faith of a Physicist* (Augsburg Fortress Publishers, 1996).

John C. Polkinghorne, *Belief in God in an Age of Science* (Yale Univ. Press, 2003).

John C. Polkinghorne, *Science and Providence: God's Interaction with the World* (Templeton Press, 2005).

John C. Polkinghorne, *Quarks, Chaos & Christianity: Questions to Science And Religion* (Crossroad Pub. Co., revised edition, 2006).

John C. Polkinghorne, *One World: The Interaction of Science and Theology* (Templeton Press, 2007).

John C. Polkinghorne, *Exploring Reality: The Intertwining of Science and Religion* (Yale Univ. Press, 2007).

John C. Polkinghorne, *Quantum Physics and Theology: An Unexpected Kinship* (Yale Univ. Press, 2008).

John C. Polkinghorne and Nicholas Beale, *Questions of Truth: Fifty-one Responses to Questions About God, Science, and Belief* (Westminster John Knox, 2009).

John C. Polkinghorne, *Theology in the Context of Science* (Yale Univ. Press, 2010).

Bernard Ramm, *The Christian View of Science and Scripture* (Eerdmans Pub. Co., 1978; originally 1954).

Del Ratzsch, *Philosophy of Science: the Natural Sciences in Christian Perspective* (InterVarsity, 1986).

Jeffrey Burton Russell, *Inventing the Flat Earth: Columbus and Modern Historians* (Praeger Paperback, 1997).

Samuel Sambursky, *The Physical World of Late Antiquity* (Princeton Univ. Press, 1988).

Michael H. Shank, editor, *The Scientific Enterprise in Antiquity and Middle Ages* (Univ. of Chicago Press, 1996).

Rudolf Simek, *Heaven and Earth in the Middle Ages: The Physical World Before Columbus* (Boydell Press, 1997).

Rodney Stark, *For the Glory of God: How Monotheism Led to Reformations, Science, Witch-Hunts, and the End of Slavery* (Princeton University Press, 2003).

G. Tanzella-Nitti, A. Strumia and P. Larrey, editors,

*Interdisciplinary Encyclopedia of Religion and Science* (online; updated monthly)

Andrew D. White, *A History of the Warfare of Science With Theology in Christendom* (New York: George Braziller, 1955; originally 1895).

Alfred North Whitehead, *Science and the Modern World* (New York: Macmillan, 1925; rep. Free Press, 1997).

Thomas E. Woods Jr., *How the Catholic Church Built Western Civilization* (Regnery Pub., 2005)

**Internet Sources (Articles)**

Chris Armstrong, Christian fathers of the scientific revolution.

Dave Armstrong, Objections to Some Atheist / Agnostic "Proof Texts" of an Alleged Flat-Earth Biblical Cosmology (vs. Ed Babinski).

Dave Armstrong, Dialogue on Biblical Cosmology, Round One (vs. Matthew Green).

Dave Armstrong, Surveys of Current Religious Beliefs of Scientists.

Dave Armstrong, Must Christian Beliefs be Falsifiable in Order to Be Rationally Held? Positivist Myths and Fallacies Debunked by Philosophers and Mathematicians

Dave Armstrong, Philosophy of Science and the Impossibility of Epistemological "Neutrality" (Especially Within Materialist or Logical Positivist Presuppositional Frameworks)

Dave Armstrong, Dialogue With an Atheist on the Relationship of Christianity and Metaphysics to the Scientific Method (vs. Sue Strandberg)

Dave Armstrong, Reply to Atheist Scientist Jerry Coyne's Position That Science & Religion Are Utterly Incompatible: Rationally Disallowing Even Theistic Evolutionism

Dave Armstrong, Old Habits Die Hard: The Garden Variety Atheist Fairy Tale of "Christianity vs. Science and Reason" Redux (vs. "drunkentune")

Hamidreza Ayatollahy, The Religious Context of Scientific Development (PDF).

Roger Bacon, On Experimental Science (from the year 1268).

Kenneth A. Boyce, Do Science and Christianity Conflict?

John Hedley Brooke, Science and Religion: Lessons From History? (*Science*, Vol. 282, no., 5386, 11 December 1998)

Michael Bumbulis, Christianity and the Birth of Science.

*Cosmos-Liturgy-Sex* website, The Christian Origins of Modern Science.

Donald DeMarco, The Christian Roots of Modern Science.

John Easton, Survey on physicians' religious beliefs shows majority faithful, *The University of Chicago Chronicle*, 14 July 2005 (Vol. 24, No. 19).

Elaine Howard Ecklund, How Religious People Misunderstand Scientists, *Science + Religion Today* website (14 August 2009).

Elaine Howard Ecklund, How Scientists Misunderstand Religious People, *Science + Religion Today* website (21 October 2009).

Elaine Howard Ecklund, What Scientists Think About Religion,

*Huffington Post*, 28 June 2010.

Elaine Howard Ecklund, Is That a Scientist in the Pew Next to You?, Baker Institute Blog (30 March 2010).

Elaine Howard Ecklund, Scientists in the Pews, *The Washington Post* (7 April 2010).

Elaine Howard Ecklund and C. P. Scheitle, Religion Among Academic Scientists: Distinctions, Disciplines, and Demographics, *Social Problems* 54: 289–307, (2007).

Elaine Howard Ecklund, Jerry Z. Park, Predicting Conflict Between Religion and Science Among Academic Scientists, *Journal for the Scientific Study of Religion* 48:2, 276-292 (June 2009).

Elaine Howard Ecklund, Jerry Z. Park, and Phil Todd Veliz, 2008. Secularization and Religious Change among Elite Scientists: A Cross-Cohort Comparison, *Social Forces*, 86(4): 1805-1840.

Robert C. Fay, Science and Christian Faith: Conflict or Cooperation?

Greg Grooms, Science and World View.

N. Gross, S. Simmons, "The Religiosity of American College and University Professors," *Sociology of Religion* 70:2, 101-129 (June 2009).

Loren Haarsma, Christianity as a Foundation for Science (PDF).

Loren Haarsma, Chance from a Theistic Perspective.

James Hannam, Christianity and the Rise of Science.

James Hannam, The Myth of the Flat Earth. ["What can be stated categorically was that a flat Earth was at no time ever an element of Christian doctrine and that no one was ever persecuted or pressured into believing it. . . . all educated people in the Middle Ages were well aware the Earth was a sphere"]

James Hannam, The Mythical Conflict Between Science and Religion.

James Hannam, Copernicus and his Revolutions.

James Hannam, Medieval Science, the Church and Universities.

Brian W. Harrison, Bomb-Shelter Theology, *Living Tradition* (May 1994).

Otto J. Helweg, Scientific Facts and Christian Faith: How Are They Compatible?, *USA Today*, March 1997.

Stanley L. Jaki, Science: From the Womb of Religion (PDF).

Pope John Paul II, Truth Cannot Contradict Truth: Address to the Pontifical Academy of Sciences (October 22, 1996).

Donald H. Kobe, Luther and Science.

Robert C. Koons, Science and Theism: Concord, not Conflict (PDF).

Erwin Laszlo, A Meeting Place for Religion and Science, *Huffington Post*, 22 June 2010.

Arnold V. Lesikar, Some of Einstein's Writings on Science and Religion.

Barton Paul Levenson, How Christianity Created Modern Science.

David C. Lindberg, The Christian Face of the Scientific

Revolution: Christian History Interview - Natural Adversaries?

David C. Lindberg and Ronald L. Numbers, Beyond War and Peace: A Reappraisal of the Encounter between Christianity and Science, *Perspectives on Science and Christian Faith* 39.3:140-149 (September 1987).

Jacques Maritain, God and Science.

John F. McCarthy, Not the Real Genesis 1, *Living Tradition* (March 1994).

John F. McCarthy, The Myth of the Self-Made Universe, *Living Tradition* (March 2006).

Sara Joan Miles, From Being to Becoming: Science and Theology in the Eighteenth Century, *Perspectives on Science and Christian Faith*, 43 (December 1991).

J. P. Moreland, Complementarity, Agency Theory, and the God-of-the-Gaps, *Perspectives on Science and Christian Faith* 49 (March 1997).

John Millam, "Christianity and the Origin of Modern Science" (.doc / html).

J. P. Moreland, Is Science a Help or Threat to Faith?

George L. Murphy, Possible Influences of Biblical Beliefs Upon Physics, *Perspectives on Science and Christian Faith* 48:2 (June 1996).

George L. Murphy, Reading God's Two Books (PDF), *Perspectives on Science and Christian Faith 58* (March 2006).

Paul Newall, The Galileo Affair (+ parts two / three / four / five; 2005).

Paul Newall, Galilean Myths.

Robert C. O'Connor, Science on Trial: Exploring the Rationality of Methodological Naturalism, *Perspectives on Science and Christian Faith* 49 (March 1997).

Ted Peters, Theology and Science: Where Are We?

The Pew Forum on Religion and Public Life, Religion and Science: Conflict or Harmony? (4 May 2009).

The Pew Forum on Religion and Public Life, Scientists and Belief (5 November 2009).

Alvin Plantinga, Religion and Science, *Stanford Encyclopedia of Philosophy*.

Alvin Plantinga, Methodological Naturalism?, *Perspectives on Science and Christian Faith* 49 (September 1997).

John Polkinghorne, Religion in an Age of Science.

John Polkinghorne, God's Action in the World.

Bernard Ramm, The Bible and Science: The Relation of Science, Factual Statements and the Doctrine of Biblical Inerrancy, *Journal of the American Scientific Affiliation*, 21 (December 1969).

Colin Russel, Without a Memory, *Perspectives on Science and Christian Faith*, 45 (March 1993).

Jeffrey Burton Russell, The Myth of the Flat Earth (1997).

Carolyn Scearce, Adelard's Questions and Ockham's Razor: Connections Between Medieval Philosophy and Modern Science.

Henry F. Schaefer III, Stephen Hawking, The Big Bang, and God.

Robert J. Schneider, "Does the Bible Teach a Spherical Earth?, *Perspectives on Science and Christian Faith*, 53 (September 2001).

Paul Seely, Reading Modern Science Into Scripture (PDF), *Perspectives on Science and Christian Faith*, 59 (March 2007).

Eric V. Snow, Christianity: A Cause of Modern Science?

Joseph L. Spradley, Changing Views of Science and Scripture: Bernard Ramm and the ASA.

*Stanford Encyclopedia of Philosophy*, Medieval Theories of Causation.

Rodney Stark, Catholicism and Science (PDF).

Rodney Stark, False Conflict: Christianity is Not Only Compatible With Science--It Created It (PDF).

Rodney Stark, How Christianity (and Capitalism) Led to Science (PDF).

G. Tanzella-Nitti, The Two Books Prior to the Scientific Revolution (PDF), *Perspectives on Science and Christian Faith*, 57 (2005).

Paul Theerman, James Clerk Maxwell and Religion, *American Journal of Physics* (April 1986).

Dick Tripp, The Complementary Nature of Science and Christianity.

Wikipedia, Age of the Earth.

Wikipedia, Antipodes.

Wikipedia, Byzantine Science.

Wikipedia, Catholic Church and Science.

Wikipedia, Christianity and Science.

Wikipedia, Flat Earth.

Wikipedia, Heliocentrism.

Wikipedia, History of Astronomy.

Wikipedia, History of Physics.

Wikipedia, History of Science.

Wikipedia, History of Scientific Method.

Wikipedia, Philosophy and Science.

Wikipedia, Relationship between religion and science.

Wikipedia, Roman Catholicism and Science.

Wikipedia, Science in the Middle Ages.

Wikipedia, Scientific Revolution.

Wikipedia, Spherical Earth.

Wikipedia, Timeline of the History of Scientific Method.

Thomas E. Woods, Jr., How the Catholic Church Built Western Civilization.

Ken Yeh, Reclaiming the Christian Roots of Modern Science.

# Lists and Overviews of Christian Scientists

List of Christian Thinkers in Science (Wikipedia)

Scientists of the Christian Faith -- Alphabetical Index (over 1600 Christian Scientists; compiled by J. P. Holding): [A] [B] [C] [D] [E] [F] [G] [H] [I] [J] [K] [L] [M] [N] [O] [P-Q] [R] [S] [T] [U-V] [W] [X, Y, Z]

The Galileo Project: Catalog of the Scientific Community in the 16th and 17th Centuries (collection of 631 detailed biographies, by Richard S. Wetstfall)

Scientists of the Christian Faith: From the Era of Galileo (522 scientists who were Christians: linked to The Galileo Project database)

Scientists and Their Gods (Henry F. Schaefer III)

Christian Influences in the Sciences (Dan Graves)

Fathers of Science (Matthew E. Bunson)

List of Jesuit Scientists (Wikipedia)

The 35 Lunar Craters Named to Honor Jesuit Scientists

Jesuits and the Sciences, 1540-1995

Adventures of Some Early Jesuit Scientists (Joseph F. MacDonnell, S. J.)

*Jesuit Geometers* (online book by Joseph F. MacDonnell, S. J.)

Seismology, The Jesuit Science

Roman Catholic Scientist-Clerics (Wikipedia)

Scientists of the Christian Faith: A Presentation of the Pioneers, Practitioners and Supporters of Modern Science (compiled by W. R. Miller)

Curricula: Science (*Catholic Encyclopedia*: links to relevant articles)

Archaeologists of the Christian Faith (W. R. Miller)

Science and Faith (many links; Arnold Neumaier)

List of Byzantine Scientists (Wikipedia)

Significant Scots: Scots Pioneers in Medicine: A Cornucopia of Pharmacopeia (George W. Rutler)

## Internet Sources (Websites)

The Faraday Institute for Science and Religion

Christians in Science

Ian Ramsey Center for Science and Religion (Oxford)

The European Society for the Study of Science And Theology

Philosophy, Science, and Christianity (web page: Dave Armstrong)

Reasons to Believe (Dr. Hugh Ross)

The Technotheology Project (W. J. Laudeman)

Christian Faith and Science (Loren Haarsma)

Gifford Lectures (University of Edinburgh)

The Center for Theology and the Natural Sciences

Zygon Center for Religion and Science

John Templeton Foundation

Metanexus Institute

Fellowship of Scientists

The Fellowship of Christian Optometrists

Christian Medical and Dental Associations

Association of Christians in the Mathematical Sciences

Christian Association of Stellar Explorers

The John Ray Initiative: connecting Environment, Science and Christianity

Society, Religion and Technology Project

The Society of Ordained Scientists

American Scientific Affiliation: A Fellowship of Christians in Science

The International Society for Science and Religion

Vatican Observatory

Counterbalance: New Views on Complex Issues

www.ingramcontent.com/pod-product-compliance
Lightning Source LLC
Chambersburg PA
CBHW031824170526
45157CB00001B/181